Engineering Management

ENGINEERING MANAGEMENT

Concepts, Procedures and Models

B. S. Dhillon, Ph.D., P.E.

Professor, Engineering Management Program
Department of Mechanical Engineering
University of Ottawa

TECHNOMIC
PUBLISHING CO., INC.

LANCASTER · BASEL

Published in the Western Hemisphere by
Technomic Publishing Company, Inc.
851 New Holland Avenue
Box 3535
Lancaster, Pennsylvania 17604 U.S.A.

Distributed in the Rest of the World by
Technomic Publishing AG

Printed in the United States of America
10 9 8 7 6 5 4 3 2

Main entry under title:
  Engineering Management: Concepts, Procedures and Models

A Technomic Publishing Company book
Bibliography: p.
Includes index p. 349

Library of Congress Card No. 87-71933
ISBN No. 87762-532-8

Table of Contents

v

Preface

In recent years interest in engineering management has been growing at a significant rate. This trend is clearly evidenced by the increase in published literature and the university programs on the subject. For example, in 1950 there was only one graduate level program in engineering management in the United States, but this number grew to six, sixteen and seventy in 1960, 1970 and 1980, respectively. In addition, today there are at least three professional journals which are specifically concerned with the discipline of enginering management.

A very high percentage of engineers move into supervisory positions between 3 and 7 years after their graduation which requires a considerable knowledge of management. So far to the author's knowledge only a few books have been written on the subject in the recent years. In addition, all of these books emphasize on the general management rather than on technical aspects of engineering management.

On the subject of general management, a vast number of textbooks are already available in the market. Therefore, an engineer requiring the knowledge of general management concepts can easily filter through these books which are written by some of the outstanding experts of management. Nevertheless, an engineer and others performing his or her job in the age of high technology not only have to have a knowledge of general management concepts but a vast knowledge of technical management concepts as well. These concepts are virtually unknown in the general management textbooks.

Basically, this book is based on author's course notes used to teach a large number of senior engineering undergraduate and graduate students over the past years. Most of the graduate students were senior practicing engineers from Department of National Defense, Chalk River Nuclear Laboratories and various other industrial organizations. Thus the material contained in this book was modified at various occasions after receiving the feedback information from both undergraduates and practicing engineers. Therefore, the emphasis of this book is on the technical aspect of the management. This is the area where most of the supervisory engineers are concerned with in their day to day operations.

For the first time to the author's knowledge this book is written to integrate the diverse areas of the technical and general management into a single volume. Topics of the book are of day to day interest to engineers and others, and are treated in such a manner that no prior knowledge is needed to negotiate contents.

The book is divided into twenty chapters.

Chapter 1 briefly discusses the various aspects of management such as history, management definition, functions of management, managerial goals and skills, management by objectives (MBO), levels of management, characteristic of management, useful information on engineering management and need for engineering management. Organizing is the subject of Chapter 2. Thus, this chapter probes into the organizational aspects of the management. The topics covered are the useful guidelines for planning an organization, design of an organizational structure, fundamental relationships in organizational structures, span of control, delegation, centralization and decentralization of organizations, methods of organization and functions of an engineering department.

Chapter 3 concentrates on human element in engineering management. Thus it discusses the needs of an engineer, routes open to an engineer for managerial positions, transition from engineer to a managerial position, activities and qualities of a manager, hints to relieve tensions, motivating engineering manpower, staff meetings, the committees and displacing managers.

Creativity is the theme of Chapter 4. This chapter covers the selective factors in creativity, creative problem solving steps, ways to develop creativity, characteristics of creative engineers and managers, climate for creativity, attributes of a manager of creative people, barriers to creative thinking; generation, presentation and evaluation of new ideas; ways to kill ideas and techniques for creativity.

Chapter 5 is concerned with manpower planning and control. Four mathematical models which will directly or indirectly aid in making useful manpower planning and control decisions are presented. A list of selected references is presented at the end of the chapter.

Chapter 6 considers the important topic of selecting engineering projects. The areas such as project selection factors, procedures for engineering project selection, feasibility analysis and project selection models are covered.

Chapter 7 studies the various aspects of project management. The topics discussed in this chapter are the need for project management, characteristics of a project management procedure, responsibilities of a project organization, actions to stimulate project success, the project manager, critical path scheduling techniques and advantages and disadvantages of the critical path method.

Management of technical proposals and specifications is the theme of Chapter 8. The chapter describes the types of technical proposals, upper management considerations in the development of a proposal, a procedure to prepare effective engineering proposals, format of a proposal, customer relations in proposal preparation, engineering specification classifications, hints for writing specifications, specification layout, military specification documents, and advantages and disadvantages of engineering specifications.

Chapter 9 explores the vital topic of management of engineering contracts. The topics such as essential provisions of a contract, contract documents, classifications of contracts, selecting a contractor for a project, types of tender, determining the progress of a contract, contract negotiation procedure, attributes of a negotiator, bids and formulas for determining escalation in price are briefly described.

Chapter 10 consists of techniques to make engineering management decisions. These techniques are the optimization approaches, business operation analysis, forecasting models, discounted cash flow analysis, depreciation methods, decision trees and fault trees. The chapter contains 22 examples with their solutions.

Mathematical models for engineering management decision making are covered in Chapter 11. Thus various mathematical models for large plant investment, general investment and equipment repair facility are presented.

Another important topic which usually interests every engineer is product developing and costing. This is described in Chapter 12. The chapter is divided into two parts, i.e., product developing and product costing. Under the product developing the topics such as reasons for developing new products, tasks to manage the development of new products, product manager and causes of product failures are discussed. The product costing aspect of the chapter covers topics such as reasons for product costing, system cost estimation procedure, life cycle costing, cost-effectiveness analysis, new product pricing and maintenance cost formulas.

Chapters 13 and 14 are concerned with management of engineering design and engineering drawings, respectively. Both these topics are explored in a fair depth.

Value engineering and configuration management are briefly explored in Chapter 15. Both topics are discussed separately in the chapter.

Chapter 16 studies the management and product assurance sciences. This chapter covers basics of reliability, quality control, system safety and maintainability management.

Engineering maintenance management is covered in Chapter 17. The chapter summarizes procedures and models to manage the maintenance function.

Marketing is the theme of Chapter 18. Various important aspects of marketing along with several mathematical models are presented.

Chapter 19 is concerned with product warranties and liabilities. Important aspects of these two topics are explained.

Chapter 20 describes the topic of work study. The chapter briefly covers both the managerial and technical aspects of the work study topic.

The book is intended for the following people:

(i)   Senior undergraduate and graduate students of industrial engineering and engineering management, respectively
(ii) Practicing engineers, managers and others
(iii) Senior undergraduate and graduate students of engineering and general management

Some of the chapters of the book may be of interest more to one group of people than another. Therefore the instructor has to use his or her judgement to select chapters of the book for the teaching purposes by taking into consideration the background of the students, their interest, outline of the course and so on. Each chapter consists of lists of exercises and source references. In all the chapters of the book, whenever a new concept is introduced usually the source reference is given. In addition, throughout the book whenever a mathematical concept is introduced, it is usually supported by example(s) along with their solutions. The book contains about 70 examples.

I would like to thank my friends, former students and leading professionals who, through discussions, have influenced my thoughts on various areas of this text. I would also like to thank Dr. S. N. Rayapati for preparing diagrams for this book. I wish to express my thanks to my family, relatives and friends for their interest and constant encouragement. In particular, I am grateful to my relatives Dr. R. S. Grewal, Mr. J. S. Grewal and Mrs. Raj Grewal for their encouragement at the moment of need. And last but not least, I thank my wife, Rosy, for her patience and tolerance as well as for typing the entire manuscript.

B. S. DHILLON

Ottawa, Ontario
May 1987

About the Author

Dr. Dhillon is a full professor of Engineering Management in the Department of Mechanical Engineering, University of Ottawa, and teaches engineering management courses to engineering undergraduate and graduate students and professionals. He has several years of industrial experience in business administration and design.

Dr. Dhillon attended the University of Wales where he received a B.Sc. in Electrical and Electronic Engineering and M.Sc. in Industrial and Systems Engineering. He received the Ph.D. in Industrial Engineering from the University of Windsor. He is advisory Editor of *Microelectronics and Reliability: An International Journal*, and Associate Editor of *International Journal of Energy Systems*. He served as an associate editor of the *10th–13th Annual Modeling and Simulation Proceedings*, Pittsburgh, U.S.A. Dr. Dhillon is on the Editorial Board of *International Journal of Reliability Engineering and Safety*. He has published over 180 articles as well as 10 books on various aspects of Engineering Reliability and reliability and maintainability management. Some of his books are translated or accepted for translation into Russian, Chinese, German, etc. He serves as a referee to many national and international journals, book publishers and other bodies. He has presented keynote and invited lectures at various national and international conferences. Recently he served as General Chairman of two international conferences on Reliability and Quality Control held in Paris and Los Angeles.

He is recipient of the Society of Reliability Engineeers' Merit Award and the American society for Quality Control's Austin J. Bonis Education Award. Professor Dhillon is a registered Professional Engineer in Ontario, and is listed in the *American Men and Women of Science*, *Dictionary of International Biography*, *Men of Achievement*, *Who's Who in International Intellectuals* and *Who's Who in Technology Today*, etc.

CHAPTER 1

Introduction

Although some requirements for management have existed for hundreds of years, prior to the last century, the need for sophisticated management existed only in limited places. However, in today's environment business is so dynamic that one is constantly dealing with changing situations as compared to the past, for example, the replacement of the older products and older production and marketing methods with the new ones. These new products and methods are accepted only after examining various alternatives.

According to experts, probably the major reasons for the business economy change are as follows:

(i) Spending on national defense and space exploration
(ii) Spending on products by the middle income group for comfort and leisure

These two reasons have led to more demand for civilian goods and military and space materials. Furthermore, these have helped to create a situation such that one may say that our present-day business exists in a changing and constantly evolving economy. Such economic conditions have created a challenging problem to management of industrial and other organizations. Thus the pressing requirement of today's management is to tackle such a problem effectively.

This chapter briefly discusses the various introductory aspects of management and leads into the discipline of engineering management.

1.1 A BRIEF HISTORY OF MANAGEMENT

This section briefly outlines the major milestones of the management movement.

The beginning of modern management goes back to the Industrial Revolution. Basically the Industrial Revolution was the result of the development of the steam engine by James Watt in the eighteenth century. Shortly after its invention in Great Britain, it was introduced in the United States. This led the way to building railroads; however, it was not until 1830 that the first railroad was built in the United States [1]. By the mid 1850s the railroads became the first industry in the United States whose operations extended outside the boundaries of local areas. This created a challenging management problem to the railroad industry managers. Thus they were the first people who required a sophisticated approach to management.

Another factor which played an important role in the American Industrial Revolution was the invention of the telegraph by Samuel F. B. Morse in 1844.

During the last decade of the nineteenth century, men such as Cornelius Vanderbilt, Andrew Carnegie and John D. Rockefeller took advantage of the rapid transportation provided by the railroads and built their big corporations. This was another factor which created the need for a sophisticated approach to management.

Men such as Frederick W. Taylor and Henri Fayol are regarded as the fathers of scientific management and the principles of management, respectively. Frederick W. Taylor joined Midvale Steel Company after the completion of his apprenticeship in 1878. His first job with the company was as a laborer. However, within six years he became the chief engineer. In 1895 Taylor presented his views on management in the *Transactions of American Society for Mechanical Engineers* [2]. Basically, his views were concerned with assigning the right worker for each job and finding the most suitable method to perform a job.

On the other hand, Henri Fayol, a Frenchman and an engineer, outlined the 14 principles of management: for example, order, initiative, division of work, centralization, discipline and line of authority. In 1916, he published a book [3,4] entitled *Administration Industrielle et Generale*. This is regarded as a major document of Fayol's thoughts. It was first translated into English by J. A. Coubrogh in 1930 and later by Constance Storrs in 1949.

The first conference on the subject of scientific management was held in October, 1911 [1]. In the period from 1910 to 1940 various professional societies related to the management field were formed. For example, the American Management Association, Society of Industrial Engineers, Society to Promote the Science of Management, and Society for Advancement of Management were formed in 1923, 1917, 1912, and 1936, respectively. According to Reference [5] most engineering schools in the United States were also teaching classes on the subject of management by the year 1925.

Another important milestone in the history of management occurred in 1924 when a study on human relations was started by the National Research Council of the National Academy of Sciences. This project is known as the Hawthorne study which was carried out at the Hawthorne Plant of Western Electric in the state of Illinois. During the course of this study various experiments were conducted, for example, investigating the effect of varying illumination, length of work day and rest periods on productivity.

After the translation of Henri Fayol's book into English, there was a general agreement by 1955 that the topic of management should be taught on similar lines specified by Fayol. However, this general agreement did not last very long and the period of fragmentation followed which lasted into the early sixties. In the late sixties the systems approach to management was emphasized. However, in the seventies, the contingency approach was thought to be an answer to the management problem. This approach states that different management procedures or approaches are needed to handle different situations and conditions.

According to Reference [1], the recent trend is towards the cost-saving methods. Finally, it may be added that engineers have played an important role in the development of the management discipline.

1.2   DEFINITION OF MANAGEMENT

There are various definitions which are used to describe management. According to Reference [1] even today there is no single definition of management which is accepted by everyone. However, for our purpose we will define management as follows [1].

"It is a process of work that involves guiding a class of persons to accomplish stated organizational objectives." The meaning of management is described graphically [6] in Figure 1.1. The diagram shown in Figure 1.1 is divided into three parts. These parts are the three distinct elements of the management definition. Thus parts I, II and III contain basic resources (i.e., people, money, etc.), functions of management (i.e., organizing, planning, staffing, motivating and controlling) and organizational objectives, respectively.

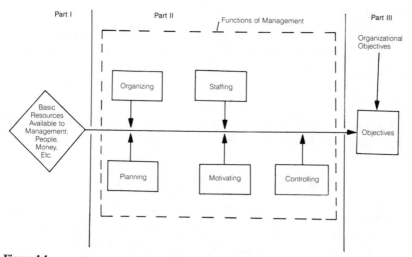

Figure 1.1.
Graphical description of the management definition.

1.3   THE FUNCTIONS OF MANAGEMENT

The basic functions of management are specified in Figure 1.1 (part II). These functions are as follows:

(i) Planning

(ii) Organizing

(iii) Staffing

(iv) Motivating

(v) Controlling

All of these functions are briefly described below [7].

PLANNING

This involves activities such as determining rules and procedures, setting objectives, scheduling and forecasting materials.

ORGANIZING

This is concerned with grouping and assigning activities, delegating authority to subordinates, and so on.

STAFFING

This involves activities such as selecting, hiring and training employees, determining the manpower needed, and setting standards to measure employees' performance.

MOTIVATING

This is concerned with guiding employees to meet performance standards, rewarding employees, maintaining morale, and outlining the objectives to employees.

CONTROLLING

This involves activities such as monitoring the actual performance, comparing to set standards and taking corrective measures.

1.4   TYPES OF MANAGERIAL SKILLS

A manager is expected to possess six types [8] of managerial skills in order to perform his or her tasks effectively. The amount of utilization of such skills varies according to the level of the manager on the management ladder. For example, the need for the technical skill will be greater at the lower level of the management ladder than at the top management level. However, the middle management positions require a fair use of all six skills. The types of such skills are as follows:

  (i) *Human Skill:* It is concerned with the ability of a manager to get along with other people. This type of skill is essential to all managers, irrespective of their rank on the management ladder. However, its requirement is greater at the low level management.

 (ii) *Conceptual Skill:* It is concerned with the manager's ability to comprehend the role of each department or group in the organizational structure and the complexities of the entire organization. The requirement for this skill is much greater at the top end of the management structure than at the lower end.

(iii) *Technical Skill:* It is concerned with the manager's ability to use certain procedures and knowledge in performing his or her work duties. The greatest need for this skill is in the low level management positions. At the top management level, it is the least required.

(iv) *Decision-Making Ability:* It is self-explanatory and is needed more in top level management positions than in the low level positions.

 (v) *Analytical Skill:* It is concerned with the manager's ability to use analytical methods when analysing work problems. The requirement for this skill is almost the same in all levels of management.

(vi) *Communication Skill:* It is concerned with the manager's ability to provide information to other people in both written form and orally. This skill is equally important in all levels of the management ladder.

1.5 MANAGEMENT BY OBJECTIVES (MBO)

This is an approach used to set objectives. Both superiors and subordinates participate in setting objectives. Therefore the MBO approach may simply be described as a management philosophy whose basis is to convert the goals of the company into personal goals.

This technique has many applications. For example it serves [7] as a

  (i) Planning technique

 (ii) Motivation technique

(iii) Control technique

The management by objective process is composed of various steps. These are concerned with establishing organizational, department and individual goals and feedback of information. Once the organizational goals are set then the department managers and their superiors jointly set the departmental goals.

After setting the departmental goals, the department managers discuss such goals with their department members and then these members are asked to develop goals for themselves. Once each individual has developed his or her goals, the department manager and the individual jointly set the goals for the latter and the time period to acommplish set goals. The feedback information on set goals is received through review meetings between superiors and subordinates.

According to Reference [7], the following conditions are necessary for the management by objectives success:

(i)   Concise and timely information feedback to employees. According to Reference [9], there should be four such information feedback sessions per year.

(ii) Commitment of the management

(iii) Adequacy in time and resources

(iv) Clarity of goals

1.6   GOALS OF MANAGERS

There are various goals which are pursued by managers. According to a survey reported in Reference [10], the managers see several goals important to them. Some of them are as follows:

(i)   Welfare of employees

(ii) Profitability and productivity

(iii) Organizational stability and efficiency

(iv) Growth and industrial leadership of the organization

1.7   MANAGEMENT LEVELS

According to Reference [11], a management structure usually has three levels of management. These management levels are as follows:

(i)   *Top Management Level:* This is the highest level of the management structure and is composed of management personnel such as president, vice-presidents, etc. The basic function of these people is to set organizational policy.

(ii) *Middle Management Level:* This level of management falls between top and low management levels. Furthermore, it is composed of management personnel such as general managers, departmental managers, etc. Basically the function of this level of management is to prepare intermediate plans so that the policies of the upper management are implemented. In other words the basic function is the execution of policies developed by the top management.

(iii) *Low-level Management:* This is the lowest level of the management structure and is composed of management personnel such as supervisors, section mangers and foremen. These persons look after the day-to-day operations of the local scene. Furthermore, they develop and implement short-range plans concerning day-to-day operations.

1.8 MANAGEMENT CHARACTERISTICS

This section lists selective characteristics of the management [6]. These are as follows:

 (i) Management is not people; it is an activity. This activity is performed by people, for example, executives, managers and supervisors. Thus an activity can be studied.
 (ii) Management has a purpose because it is practiced to accomplish a goal.
(iii) Management requires experience, skill and specific knowledge for its effective use.
(iv) Management is an unseen force.
 (v) Broadly speaking, management is associated with the efforts of a group of persons. However, it can also be associated with efforts of a single person.
(vi) Computers help management but do not replace it.
(vii) Management is an important means for making things happen.

1.9 USEFUL INFORMATION ON ENGINEERING MANAGEMENT

The field of engineering management has been gaining prominence in recent years. However, it is not that new; it has been with us for a number of decades. For example, according to Reference [12], in 1950 there was one graduate level program in engineering management in the United States. This number grew to 6, 16, and 70 in 1960, 1970, and 1980, respectively. Presently various professional societies are concerned with promoting the field of engineering management. These are as follows:

 (i) American Society of Engineering Management (ASEM)
 (ii) Institution of Electrical and Electronics Engineers (IEEE). It publishes two journals on Engineering Management.
(iii) American Society for Engineering Education (ASEE). One of its divisions is concerned with engineering management.

(iv) Institute of Management Sciences (TIMS). One of its divisions is also promoting the field of engineering management.

1.9.1 Engineering Management Journals

To the author's knowledge only the following journals are specifically devoted to engineering management:

(i) *IEEE Transactions on Engineering Management*

(ii) *IEEE Engineering Management Review*

(iii) *Engineering Management International*, published by Elsevier Science Publishers B.V., Amsterdam, The Netherlands

ASCE

1.9.2 Books on Engineering Management

So far there has been only a small number of books which have the words "Engineering Management" in their titles. Some of them are as follows:

(i) V. Cronstedt. *Engineering Management and Administration.* McGraw-Hill Book Company, New York, 1961.

(ii) T. G. Hicks. *Successful Engineering Management.* McGraw-Hill Book Company, New York, 1966.

(iii) D. I. Cleland and D. F. Kocaoglu. *Engineering Management.* McGraw-Hill Book Company, New York, 1981.

(iv) R. E. Shannon. *Engineering Management.* John Wiley & Sons, New York, 1980.

1.10 NEED FOR ENGINEERING MANAGEMENT

In recent times, the interest in engineering management has been growing at a significant rate. This is clearly evidenced by the increase in published literature and the university programs on the subject. There could be various factors for such an increase, for example, recent emphasis on productivity, increase in engineers' desire to perform management functions and competition for promotion.

For engineers, the management skill is an essential element to move upward on the management ladder. Furthermore, many of them move into the supervisory positions between 3 and 7 years after their graduation. Most of such engineers with a degree in the traditional engineering fields are not well prepared to take management responsibilities. At that time many of them realize the importance of management knowledge.

According to Reference [12], in the United States, 40 percent of industrial executives are from engineering backgrounds. This is another important factor which indicates the need for engineering management.

In recent years, many engineers, to gain management knowledge, attend business schools and obtain a master of business administration degree. Although this is a good route to fulfill their managerial needs, a degree in engineering management will most probably fulfill their need more closely because it emphasizes more the technical aspect of management than the general aspect.

This is what most of engineering management jobs probably require because in managerial positions usually engineers play the roles of technologist and manager. Thus engineering management is a discipline which will fulfill the need of an engineering manager because it is the intersection of engineering and management.

1.11 SUMMARY

This chapter briefly discusses the various introductory aspects of general management and engineering management. It begins by describing the need for management and then goes on to summarize the history of the management discipline. After reviewing the historical aspect of management discipline, it was concluded that engineers have played an important role in the development of management discipline. Furthermore, both the definition and functions of management are briefly discussed. Five basic functions of management are briefly described.

A manager has to possess various types of managerial skills. Thus six types of such skills are briefly discussed. These skills are human, conceptual, technical, decision-making ability, analytical and communication. Management by objectives and goals of managers are the other two topics which are briefly outlined. The next topic discussed is the levels of management. Three levels of management, i.e., top, middle and low-level management, are explained. In addition, seven characteristics of management are outlined.

The last two topics discussed are specifically concerned with engineering management. These are the "useful information on engineering management" and "need for engineering management."

1.12 EXERCISES

1. What were the basic principles of scientific management as developed by Frederick Winslow Taylor?
2. What are the differences between general management and engineering management?

3. Briefly explain the ten characteristics of management.

4. Discuss at least three different definitions of management and compare them with each other.

5. Describe the management by objectives (MBO) process.

1.13   REFERENCES

1. Rue, L. W. and L. L. Byars. *Management: Theory and Application.* Homewood, IL 60430:Richard D. Irwin, Inc. (1980).

2. Taylor, W. "A Piece-Rate System," *Transactions ASME,* 16:856–883 (1895).

3. Fayol, H. *Administration Industrielle et Generale.* The Society de l'Industrie Minerale (1916).

4. Fayol, H. *General and Industrial Management.* Sir Isaac Pitman and Sons (1949).

5. Mee, J. F. "Management Teaching in Historical Perspective," *The Southern Journal of Business,* 7:21 (May 1972).

6. Terry, G. R. *Principles of Management.* Homewood, IL 60430:Richard D. Irwin, Inc. (1972).

7. Dessler, G. *Management Fundamentals: Modern Principles and Practices.* Reston, VA:Reston Publishing Company (1982).

8. Mondy, R. Wayne, R. E. Holmes and E. B. Flippo. *Management: Concepts and Practices.* Boston:Allyn and Bacon Inc. (1980).

9. Ivancevich, J., J. A. Donnelly and H. L. Lyon. "A Study of the Impact of Management by Objectives on Perceived Need Satisfaction," *Personnel Psychology,* 23:139–151 (September 1970).

10. England, G. "Organizational Goals and Expected Behavior of American Managers," *Academy of Management Journal,* 10:107–117 (1967).

11. Kelly, J. *How Managers Manage.* Englewood Cliffs, NJ 07632:Prentice-Hall, Inc. (1980).

12. Cleland, D. I. and D. F. Kocaoglu. *Engineering Management.* New York:McGraw-Hill Book Company (1981).

CHAPTER 2

Organizing

2.1 INTRODUCTION

Ever since the beginning of the Industrial Revolution, industrial organizations have been increasing in size, number and the size of their operations. Furthermore, today's industrial companies have to operate in more competitive environments as compared to the environments of the past. These and other factors have led the way to organizing the operations of industrial companies in such a way that they function most efficiently and at the minimum cost. To organize a company which can function effectively with minimum expense requires inputs from various diverse areas. In other words, it is not that type of task which can be handled by one individual alone.

According to Reference [1], in broad terms, the organizing is the process of establishing a structure for the organization so that it helps the manpower of the organization to function systematically to fulfill the organizational goals effectively.

The main objective of this chapter is to briefly present the various important aspects of organizing which will be useful to engineers. Therefore the topics such as organizational methods, delegation, span of control and functions of an engineering department are discussed in the chapter.

2.2 USEFUL GUIDELINES FOR PLANNING AN ORGANIZATION

This section presents some useful points which will be helpful for planning an organization. The points are described below [2]:

(i) The person who delegates responsibility is accountable.

(ii) Design positions in such a way that both functional and managing duties are separate.

(iii) The design of each element of the organization must be such that it helps to fulfill its own objectives as well as that of its parent enterprise.

(iv) To obtain useful policies and decisions, the delegation of authority and responsibility must be as close as possible to the point of action.

Figure 2.1.
The sequence of steps involved in designing an organizational structure of a company.

(v) From the top to the bottom of the organization the line of responsibility and authority must be clear.

(vi) A person in the organization must receive orders from only one higher authority.

(vii) The organization should be structured so that the levels of management are minimum.

2.3   DESIGNING AN ORGANIZATIONAL STRUCTURE OF A COMPANY

The organizing is a distinct function based on strategy. Thus, according to Alfred Chandler, the structure follows strategy [3]. His well known study found that over a number of years the organizational structures of corporate giants such as General Motors, Du Pont, Sears and Standard Oil were changed due to changes in their goals and strategy. Therefore it may be said that an organizational structure is not a static structure. Nevertheless, to design such a structure involves four steps. The sequence of these steps is shown in Figure 2.1. All these steps are described briefly below:

STEP A

Divide the company into major subgroups, horizontally. These subgroups must correspond to those activities which are critical to fulfilling the objective.

STEP B

Decide whether each activity of Step A should be line or staff activity.

STEP C

This step is concerned with creating authority relationships among positions. Usually, the step involves further subdivisions of units, creation of line of command and staff units.

STEP D

This step is basically concerned with designing and delegating tasks.

2.3.1   Reasons for Having an Organizational Chart

There are various reasons to have an organizational chart [4]. Some of them are as follows:

(i)   It assigns responsibilities to individuals.
(ii)   It helps to identify poor or strong control.
(iii)   It makes the management functions simpler.
(iv)   It serves as a framework for budgeting and scheduling.
(v)   It outlines fundamental relationships.
(vi)   It helps to give sense of security.
(vii)   It serves as a basis for directives.
(viii)   It outlines basic authority.
(ix)   It helps to improve communication channels.
(x)   It serves as a reference document for various purposes.

2.4   FUNDAMENTAL RELATIONSHIPS IN ORGANIZATIONAL STRUCTURES OF COMPANIES

In organizational structures the following three relationships are used [5]:

(i)   *Line:* This is known as the line or direct authority. Diagramatically

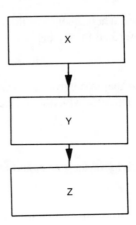

Figure 2.2.
Line authority.

this is shown in Figure 2.2. This is that type of authority which flows from superior X to subordinate Y and then through subordinate Y to subordinate Z. This way it forms a line from the top of the management ladder to the lowest organizational level. This is why the name "line authority" is given to represent such a situation.

(ii) *Group:* This is known as the group or multiple reporting. This relationship is shown in Figure 2.3. In this situation, a number of subordinates reports to a single superior. For example, in Figure 2.3, the subordinates B, C, D and E report to a superior X. Usually a larger number of individuals at the bottom end of the organizational structure than at the top report to a superior.

(iii) *Staff:* The purpose of the advisory staff is to help the line management to carry out the organizational objectives most effectively. According to Reference [1] which quotes Claude George that this concept is not new, its history traces back to Alexander the Great (336–323 B.C.). Alexander's armies were the first to employ such a concept. The staff relationship is shown in Figure 2.4. In this diagram the staff person reports to superior X and performs the advisory function which helps X.

History

2.5   SPAN OF CONTROL

According to Reference [6], the span of control problem has been there for a long time; in fact, it is as old as organization itself. This problem is simple to describe but very difficult to find a solution for. The problem basically is to

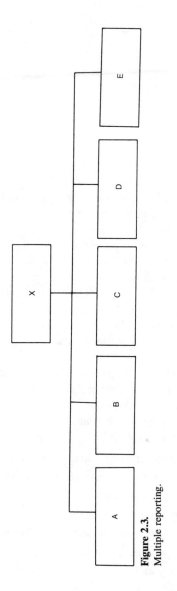

Figure 2.3.
Multiple reporting.

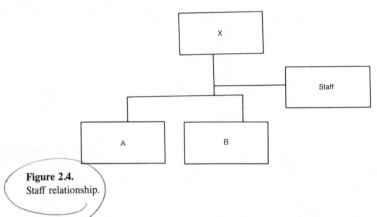

Figure 2.4.
Staff relationship.

decide how many individuals should report to a superior so he or she can manage them effectively. Therefore, each and every company has to decide the span of control at each organizational level.

Many experts have proposed various numbers of people to be supervised by a superior. For example, some people may say that at the upper levels of management the ideal number of individuals to be supervised by a superior should be between four to eight and at the lower levels between eight to fifteen or more. According to the survey of 100 large American companies conducted by the American Management Association [6], the executives reporting to presidents varied from one to 24. Furthermore, the result of this study indicated that there were only 26 presidents who were supervising six or less individuals.

According to Reference [6], the present-day trend is not to believe, in regard to span of control, that there is a numerical limit which can be applied widely, but to find out for individual cases the causes of limited span.

The following factors play an important role in deciding how many individuals should report to a manager [6]:

   (i) Ability of the manager to get along with his or her subordinates

  (ii) Capacity of the manager to comprehend quickly

 (iii) Loyalties of the subordinates

 (iv) Respect of subordinates

  (v) Frequency and time impact of superior–subordinate relationships

The last factor is the most important one. If a manager is able to reduce significantly such frequency and time impact then it may help him or her to supervise more people effectively. The following factors play an important role in shaping frequency and time impact of superior–subordinate relationships:

   (i) Communication methods' effectiveness

  (ii) Proper qualifications of individuals

 (iii) Clarity in the authority delegation

(iv) Degree of personal touch
(v) Clarity in the job plans for subordinates
(vi) Rate of changes in the organization
(vii) Degree of utilization of objective standards or other means to find
out whether the individuals make use of plans

2.5.1 Lockheed's Span of Control Model

This model was developed by the Lockheed Missile and Space Company after an extensive effort [6,7]. The model makes use of the following factors:

(i) Geographical locations of subordinates reporting to a manager
(ii) Nature of work performed and department managed
(iii) Similarity of functions performed by subordinates
(iv) Organizational help given to the manager
(v) Coordination required
(vi) Managers' and their organization units' planning, functions importance, complexity and time requirements
(vii) Degree of direction and control needed by subordinates

The company assigns a weight to each of the above variables in order to determine the optimum span of control.

2.6 DELEGATION

This is concerned with assigning responsibility and authority to subordinates by an executive. The delegation of authority and responsibility to subordinates is necessary because superiors work through others.

Delegation is not an easy task. It suffers from various managerial and subordinate obstacles. These obstacles and other areas concerning delegations are discussed below.

2.6.1 Managerial Obstacles to Delegation

In practice many managers are reluctant to delegate. William H. Newman [8] gives the following reasons for their reluctance:

(i) Managers may have poor confidence in individuals who work for them.
(ii) Ultimately the executives are responsible for the tasks performed by their subordinates; therefore, managers may be wary of risk.

(iii) Managers may have false beliefs that they can do a better job than their subordinates. Therefore they may feel that this is their duty to perform such tasks themselves.

(iv) Lack of appropriate control measures to make management aware of impending difficulty

(v) Lack of ability to direct

2.6.2   Subordinate's Obstacles to Delegation

Again according to William H. Newman the reasons for obstacles to delegation from the subordinate's side may be as follows:

(i) Lack of adequate facilities to accomplish the job properly

(ii) Lack of self-confidence

(iii) Subordinate's reluctance to sort out the problem by himself or herself because he or she finds it easier to obtain the help of the superior

(iv) No worthy incentives for subordinate's extra responsibility

(v) Fear of mistakes. The subordinate believes that more responsibility will lead to an increase in the probability of making mistakes.

(vi) Subordinate's heavy work load which may already be outside his or her limits

2.6.3   Useful Guidelines to Delegate Authority Effectively

Kelly [7] presents various guidelines which will be helpful for managers in delegating authority effectively. Some of them are as follows:

(i) Aim for the total understanding of the subordinate on the following points:
   (a) What is being delegated to him or her
   (b) His or her authority
   (c) Procedure to be used to review his or her work
   (d) The way he or she will get rewarded for his or her effort

(ii) Decompose the job in question into various separate tasks.

(iii) The superior should be aware of the fact that he or she can delegate authority but not responsibility. After all, a manager is accountable for his or her subordinate's action.

(iv) Select only those subordinates whose qualifications are matched with the tasks to be performed.

(v) The manager should make certain that individuals have been given adequate authority to perform their tasks effectively.

(vi) For each task to be performed, outline the responsibility standards.

(vii) Set up such standards for control which are clearly understood by subordinate.

(viii) Periodically monitor the progress made.

2.7 CENTRALIZATION AND DECENTRALIZATION OF ORGANIZATIONS

In the centralized organization the upper management has the authority to make the critical decisions. On the other hand, in the decentralized organizations the authority is distributed to lower levels of the management ladder. This section briefly discusses some aspects of decentralization along with advantages of centralization.

2.7.1 Determining Factors for the Decentralization of Authority

According to Reference [6], some of the factors which shape the degree of decentralization of authority are as follows:

(i) Thinking philosophy of management

(ii) Availability of properly trained managers which are needed to make decentralized decisions

(iii) Availability of control techniques which will help to monitor decisions made at the lower levels

(iv) Influences from external forces such as income tax policies, unions and governmental controls

(v) Dynamic nature of business

(vi) Policy uniformity and costs of decisions

(vii) Size and history of the company

(viii) Desire of individuals and groups for independence

2.7.2 Benefits of Centralization and Decentralization

This section briefly explores the advantages of both centralization and decentralization [1,2].

CENTRALIZATION

(i) It helps to reduce the number of undesirable decisions by less experienced executives.

(ii) It helps to reduce the need of experienced subordinate managers as they may be required in the case of a decentralized organization.

(iii) It helps to reduce the cost of staffing because fewer competent managers can handle the jobs and the requirement of staff.

(iv) It helps to bring important decision makers close to each other, which in turn simplifies the coordination of their efforts.

(v) It helps to eliminate duplication of work.

(vi) It helps to improve control over specialized functions.

(vii) It helps to make centralized staff expertise more efficient and simpler to use.

DECENTRALIZATION

(i) It helps ensure that the decisions are made by those managers who have the best experience of local conditions which are important for such decisions.

(ii) It helps to make the overall organization stronger by facilitating the personal development of individuals.

(iii) It helps to stimulate initiative and identification with the enterprise.

(iv) It helps in creating better feelings of satisfaction among competent individuals relative to centralization.

2.8   METHODS OF ORGANIZATION

There are various methods which are used to set up an organizational structure. Usage of each of these organizational methods depends upon company policy, product, location, skills of manpower and so on. Some of the organizational methods are described below [1,2,5,6].

2.8.1   Organization by Function

This is probably the most widely used method to organize company activities. This method calls for separating work according to discipline or subject, in addition to performing all similar work within the framework of one unit. A typical example of application of such a method is "an engineering manager being responsible to supervise all engineering work." Figure 2.5 shows a simplified diagram of a functional organization. According to Reference [5], this type of organizational approach is favoured by large research groups and enterprises. In addition, the companies with long-term projects also favour this type of organization.

ADVANTAGES

(i) It eliminates duplication of facilities.

(ii) It helps to permit technical specialization.

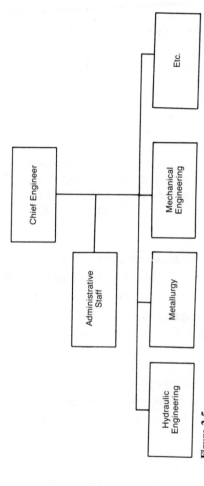

Figure 2.5.
A simplified chart of a functional organization.

(iii) It helps to make the supervision of the group easier because of group homogeneity.

(iv) It helps to distribute work load evenly.

(v) It helps to produce more uniform products.

(vi) It helps to have consistent policy.

DISADVANTAGES

(i) For cross-discipline development work this approach is not that beneficial.

(ii) It makes it difficult to shift personnel.

(iii) It slows the work flow.

2.8.2 Organization by Product

This is another approach followed by some companies to set up their organizational structure. In this case, the company is divided into divisions. Each division of the company is assigned responsibility for one particular product. A simplified chart of a product organization is shown in Figure 2.6. The advantages and disadvantages of such an organization are as follows.

ADVANTAGES

(i) It helps to develop teamwork more easily.

(ii) Profits responsibility is placed at the divisional level.

(iii) It helps to develop better coordination of functional activities.

(iv) For general managers it serves as a measurable training ground.

(v) It helps to place better attention on the product.

DISADVANTAGES

(i) It limits the contacts between persons of same specialty.

(ii) It increases the need for general managers.

(iii) It makes it difficult to maintain central services economically.

2.8.3 Organization by Territory

This is another way used to develop an organizational structure [6]. This is used in those situations where the companies are physically dispersed. This approach calls for grouping of all activities in a specified territory and then appointing a manager to look after such a group. Reasons for departmentation by territory are poor communication facilities, encouraging local participation in decision making, necessity to take prompt action and so on. A simplified chart of departmentation by territory is shown in Figure 2.7.

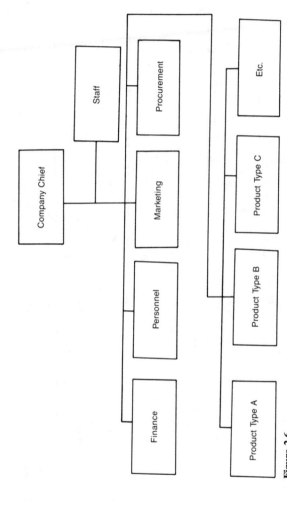

Figure 2.6.
A simplified chart of a product organization.

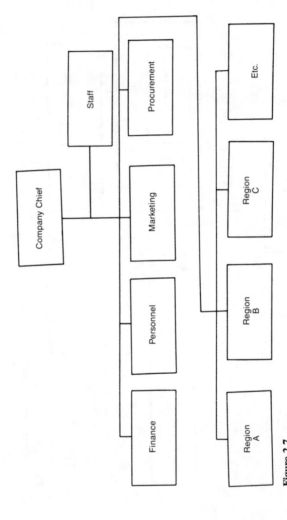

Figure 2.7.
An example of departmentation by territory.

The following are some of the advantages and disadvantages of departmentation by territory.

ADVANTAGES

(i) It is helpful to take full advantage of local operations' economies.

(ii) It helps to upgrade regional coordination.

(iii) It is useful in placing emphasis on local conditions.

(iv) It is useful in placing responsibility at a lower organizational level.

DISADVANTAGES

(i) It makes it difficult to maintain economical central services.

(ii) It increases the need for general managers.

2.8.4 Organization by Project

This is another important method of organization. Broadly speaking, it may be said that the project organization is a non-permanent organizational structure established to fulfill a certain objective. This way the most appropriate talent at the disposal of the company is grouped together to carry out a specific complex project within prescribed limits. Once the project is accomplished, the persons associated with the project are either transferred to a new project or sent back to their original or permanent department.

According to Reference [5], the small- and medium-sized enterprises tend to favour the project organizational approach. Furthermore, those companies with various short-term jobs also favour this concept of organization. Figure 2.8 shows a simplified organizational structure of a project organization.

Some of the advantages and disadvantages of the project organizational concept are as follows [2,5]:

ADVANTAGES

(i) It helps to focus attention on a single project.

(ii) It serves as a framework for team effort.

(iii) To fulfill project requirements, the people assigned to the project can be changed quickly.

(iv) It helps to obtain better coordination of large projects.

(v) It helps to encourage creativity because of the tendency to regroup people on a continuing basis and because of instable nature of the project organization.

(vi) It helps to increase the efficiency of work flow.

(vii) It serves as a useful tool for specialization by product.

(viii) It helps to make more specific the accountability and responsibility.

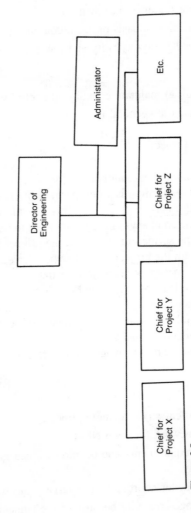

Figure 2.8.
A simple project organizational structure.

DISADVANTAGES

   (i) Inconsistent policy

  (ii) Duplication in work facilities

 (iii) Less uniform product

 (iv) Discourages technical specialization

2.8.5 Matrix Organization

Basically this is the result of combining the project and functional organizational concepts. Through combining, the greatest advantages of both approaches are obtained. This concept was originally practiced by medium-sized aerospace corporations in the fifties and sixties. The reason for its development was that such companies were not large enough to take advantage of the project organization concept.

In the matrix organization, the project manpower is on loan to the project manager. Furthermore, the project manpower reports to both the project manager and the managers of the "home" functional departments, in other words, to the managers of these departments to whom they are permanently assigned. According to Reference [9], the following are some advantages and disadvantages of the matrix organization:

ADVANTAGES

   (i) Efficient in making decisions

  (ii) Defined authority and responsibility

 (iii) Provides a better control

 (iv) Increases the role of middle management

DISADVANTAGES

  (i) It may produce surplus skills in the event of smaller projects.

 (ii) It may be misapplied to projects which do not require it.

2.9 FUNCTIONS OF AN ENGINEERING DEPARTMENT

To perform the organizing function effectively in engineering companies, the knowledge of basic functions of an engineering department is useful. Thus this section lists many of the functions of an engineering department, which, according to Reference [5], may be grouped into the following five categories:

   (i) Research and analysis

  (ii) Liaison

 (iii) Administrative

(iv) Product development

(v) Services and drafting

The functions involved in each of these categories are presented below.

RESEARCH AND ANALYSIS

The functions involved in this group are patent and literature search, search for new ideas, finding solutions to basic problems, test and evaluation, calculation and computing, analytical services, etc.

LIAISON

The functions of this category are concerned about liaison with customers, other departments, etc.

ADMINISTRATIVE

The functions such as budgeting, pricing, duty assignments, formulating policies, planning, work control, hiring, firing, training, salary review, promotion, procurement and progress reports are the components of this category.

PRODUCT DEVELOPMENT

The components of this category are preparing and analysing proposals, carrying out search for new products, standardizing parts and materials, designing new products, redesigning and so on.

SERVICES AND DRAFTING

This category is composed of functions such as release of drawings and prints, publications, model shop, drafting, etc.

2.10 SUMMARY

This chapter briefly describes the subject of organizing and its associated aspects. The chapter begins by outlining guidelines for planning an organization. Thus seven guidelines are presented. The next topics covered in the chapter are concerned with designing an organizational structure of an enterprise. A four-step approach is described to fulfill this need. In addition, ten reasons for having an organizational chart are given.

Another topic described in the chapter is concerned with the basic relationships in organizational structures. Three such relationships are explained. These are line authority, group reporting and advisory function. All three items are discussed with the aid of diagrams.

The span of control is another popular topic of interest covered in the chapter. Five factors which play an important role in deciding how many individuals should report to a manager are listed. The most important factor out of these five is known as frequency and time impact of superior–subordinate relationships. Seven factors which play a vital role in shaping the most important factor are specified. Finally, under the topic of span of control, a model developed by Lockheed Missile and Space Company for such a purpose is described. The model is based upon seven factors. All of them are briefly discussed.

The next topic explored in the chapter is delegation. Both managerial and subordinate obstacles to delegation are discussed separately. In addition, eight useful guidelines for delegating authority effectively are briefly explored.

The next topic discussed in the chapter is concerned with both centralization and decentralization of organizations. The factors which are helpful in determining the decentralization of authority are given. Advantages of both centralization and decentralization are listed.

The methods of organization is the next important topic which is covered in detail. Five different methods of organization are described. These are organization by function, organization by product, organization by territory, organization by project and the matrix organization. Four of these approaches are described with the aid of organizational charts. In addition, advantages and disadvantages of all these methods are presented.

The last topic of the chapter is concerned with functions of an engineering department. Such functions are classified into five main categories, for example, research and analysis, administrative, product development, services and drafting and liaison.

2.11 EXERCISES

1. Compare the matrix organization with the project organization.
2. Develop an organizational chart of an engineering company. Assume that this company's organizational structure is based on the organization by function concept.
3. List the advantages of having an organizational chart.
4. Describe the model developed by Lockheed Missile and Space Company for span of control with the aid of weighting factors.
5. List as many managerial obstacles to delegation as possible.
6. What are the reasons for the subordinate's obstacles to delegation?
7. How would you delegate authority effectively? Describe your approach in detail.
8. Discuss the disadvantages of the following:
   (i) Centralization
   (ii) Decentralization

9. Describe briefly the following terms:
   (i) Organizational structure of a company
   (ii) Line authority
   (iii) Departmentation
10. What are important functions of an engineering manager?

2.12 REFERENCES

1. Mescon, M. H., M. Albert and Khedouri. *Management: Individual and Organizational Effectiveness*. New York:Harper & Row Publishers (1981).
2. Karger, D. W. and R. G. Murdick. *Managing: Engineering and Research*. 200 Madison Ave., New York:Industrial Press Inc. (1969).
3. Chandler, A. P. "Strategy and Structure," in *History of the American Industrial Enterprise*. Cambridge, MA:The MIT Press (1962).
4. "Organization of the Engineering Department," in *Management Guide for Engineers and Technical Administrators*, N. P. Chironis, ed. New York:McGraw-Hill Book Company, pp. 26–39 (1969).
5. Tangerman, E. J. "Engineering Organization Charts for 25 Modern Companies," in *Management Guide for Engineers and Technical Administrators*, N. P. Chironis, ed. New York:McGraw-Hill Book Company, pp. 24–25 (1969).
6. Koontz, H., C. O'Donnell and H. Weihrich. *Management*. New York:McGraw-Hill Book Company (1980).
7. Kelly, J. *How Managers Manage*. Englewood Cliffs, NJ 07632:Prentice-Hall, Inc. (1980).
8. Newman, W. H. "Overcoming Obstacles to Effective Delegation," *Management Review*, 36–41 (January 1956).
9. Pywell, H. E. "Engineering Management in a Multiple (Second- and Third-Level)-Matrix Organization," *IEEE Transactions on Engineering Management*, Em-26:51–55 (1979).

CHAPTER 3

Human Element in Engineering Management

3.1 INTRODUCTION

The human element plays an important role in the success or failure of an engineering organization. For example, a company may have a novel and potentially profitable engineering product to offer to the market; however, poor human management within the company may make it an unprofitable product. Therefore, the management in engineering organizations not only has to come up with the most desirable product but also has to manage the human element within the company most effectively.

According to Glueck in Reference [1] the human resources of an enterprise are the most important resource. Glueck has cited the following two reasons in his statement:

(i) The human sources are a major component of cost in most enterprises. For example in chemical or petroleum plants, the cost of labor varies between 25 to 30 percent of total operating cost. It is much higher in places such as schools, universities and so on.

(ii) The human resources influence productivity of a company. For example, people set objectives of an organization, design and develop new products, operate machines, etc.

The aim of this chapter is not to make the reader an expert in the human aspect of the management but to introduce him to the various important elements of such topics. Therefore, the topics discussed in this chapter are directly or indirectly related to human aspects of the engineering or general management. These topics are concerned with areas such as the responsibilities and attributes of an engineering manager, displacing of supervisory personnel, routes open to engineers for managerial positions, needs of an engineer and motivating engineering manpower.

3.2 NEEDS OF AN ENGINEER

According to Randsepp [2], some of the needs of an engineer are as follows:

(i) Job security with respect to his or her attainments

   (ii) Proper work facilities

  (iii) Necessary technical assistance

  (iv) Stimulating and challenging work

   (v) Taking part in those decisions which will affect him or her

  (vi) Adequate supporting staff

 (vii) Bosses who are competent

(viii) Recognition for his or her work from the management

  (ix) Economic advancement

   (x) Proper work assignment

  (xi) Opportunities for self-development

 (xii) Employment with an organization which has clearly defined responsibility and authority

(xiii) Employment with a reputable organization

(xiv) Work variety

 (xv) Chance of his or her ideas being practiced

(xvi) Independence to attack a work problem

3.3 ROUTES OPEN TO AN ENGINEER FOR MANAGERIAL POSITIONS

According to Reference [2], there are various routes which are open to engineers to obtain managerial positions. Some of them are as follows:

   (i) *Changing job:* In the early stage of an engineer's career, it is advisable to look for another job elsewhere once the necessary experience is obtained over the period of two or three years at one place because generally in many companies the promotional prospects are rather slow.

  (ii) *Outstanding technical competence and ability to organize:* This is one of the best routes for those persons who are doing very well at their professional specialties to obtain managerial positions. This way such persons can seek broader technical responsibilities which will subsequently lead the way towards the managerial ranks.

  (iii) *Obtaining a graduate degree in management fields and doing a reasonably good job at the place of work:* This route is becoming popular. Surveys show that those engineers or scientists who have obtained a master's degree in management within one to three years after their first graduation earn on average 17 percent higher salaries than their counterparts with a master's degree in engineering [2].

  (iv) *Sponsor-protégé arrangement:* This type of situation exists in most organizations. Generally any manager looks for certain persons whom he likes and can depend upon. Therefore when such a manager moves up the ladder, he tries to get promotions for his closest people so they can work for him in his new job. In this situation an engineer has to make sure that he or she chooses the right manager to work for. Otherwise the end result will not be encouraging.

(v) *Completing the assigned jobs with outstanding quality and on time.*

(vi) *From consulting engineer to manager:* This way an engineer develops his specialty by becoming a consulting engineer then moving to a managerial rank. In this situation the engineer has to make sure that his or her specialty is not too specialized. Otherwise there will be a problem finding a managerial position.

(vii) *Leadership in professional activities and reasonable good service with the company:* This is another route to a managerial position. However, it is a lengthy route. The leadership in professional activities has to be obtained by writing articles, books and securing offices in the professional societies, associations, clubs and community work. This way an engineer can secure "high visibility" with the officials of his company.

3.4   FROM ENGINEER TO A MANAGERIAL POSITION

When an engineer switches from a purely technical function to a managerial position, he or she finds certain changes in responsibilities, personnel habits and so on. Therefore, an engineer has to be prepared for such changes; otherwise, he or she will find it difficult to perform his or her job effectively. This section discusses those changes which are usually experienced by an engineer performing a technical function who then switches to a managerial position [3]. Some of these changes are as follows:

(i) *He or she will be concerned with generalities:* In management an engineer has to work with generalities such as negotiation, supervision, delegation and sales instead of specifics such as pressure, force, length and weight in engineering.

(ii) *He or she will be required to make speeches:* Many of the management problems are solved in meetings. Therefore, a manager is expected to propose, explain and defend his or her ideas in such gatherings. In addition, sometimes a manager is expected to make speeches to public audiences as well. Therefore, a manager has to learn to talk and think on his or her feet for the success of his or her career.

(iii) *He or she will be concerned with the human relationships:* In management one is concerned with achieving the maximum productivity from people with minimum friction and least effort. The following human relations rules will be useful to the newcomers and others in management. These are as follows:

a) Learn to treat each person as an individual.
b) Show firmness.
c) Try to be impartial.
d) Try to be fair.
e) Give credit to others whenever it is necessary.
f) Look for what your people are able to perform, not what you are looking for.
g) Express genuine interest in the problems of other people.
h) Show genuine interest in other people.

(iv) *He or she will be required to change his or her thinking:* In the past, an engineer's thinking was confined to only one technical project. However, in management his or her thinking will be broader. The performance of the department will be more important to the engineer once in management. Furthermore, in management one's thinking will be concerned with selling to others, thinking of alternatives, thinking ahead of problems, thinking as listening to people, etc.

(v) *He or she will be required to change his or her reading habits:* Usually an engineer reads the technical magazines. However, once in the managerial position, the engineer still has to read the same technical magazines plus a few more on management. This means more time to be spent on reading. To overcome this problem, the manager looks for results rather than understanding each and every step of an approach or a process given in these magazines.

(vi) *He or she will be concerned with training people:* Due to rapid growth in modern technology nowadays, frequently people have to be trained to handle new jobs effectively. The training is one of the concerns of the manager.

(vii) *He or she has to learn how to delegate:* No manager can perform all the tasks required from his or her department. Therefore, the orders to perform such tasks have to be given to others. The new manager has to learn the secrets of giving orders properly and pleasantly.

3.5 ACTIVITIES AND QUALITIES OF A MANAGER

This section briefly discusses the functions and qualities of a manager separately.

3.5.1 Activities of a Manager

Usually a manager is concerned with eight basic activities [4]. These activities are shown in Figure 3.1.

The activities shown in Figure 3.1 are considered to be self-explanatory; therefore, they are not described. However, they are discussed in Reference [4].

3.5.2 Qualities of a Manager

A successful engineering or non-engineering manager possesses certain qualities or attributes. According to Reference [5], a good engineer should possess the following qualities:

(i) *Flexibility and fairness:* These two are the important attributes of a manager. Flexibility is necessary when one is dealing with people because of varying behaviour of humans. Behaviour of people is subject to the environments in which they perform. Fairness is another attribute that a manager should possess. A manager who has established his or her record as a fair boss may sometimes be tolerated by his or her subordinates when being totally unfair.

(ii) *Reasoning ability and tolerance:* Reasoning ability is another attribute that a manager must possess because sometimes it becomes necessary to handle a situation, such as when a subordinate believes he or she is correct but the manager sincerely feels that he or she is wrong. In this situation the manager has to cater to emotions. Tolerance is another quality that a manager should be gifted with. A manager should understand that no person is perfect. Thus the manager should be able to tolerate the shortcomings of his or her people.

(iii) *Freedom from suspicion and prejudice:* To generate confidence in the people working for the manager, mutual trust is important. The suspicion will not lead anywhere. A good manager never has prejudice against any individual from a minority group.

(iv) *Empathy and quickness to praise and criticise:* Empathy is an important attribute of a manager. Empathy can simply be described as "the capacity to feel what other persons feel." Another important attribute of a manager is to praise quickly the persons who have performed a good job. However, in a situation where the job was not properly accomplished, the manager must also act quickly to criticise constructively those involved with that job. A good manager avoids off-the-cuff criticism.

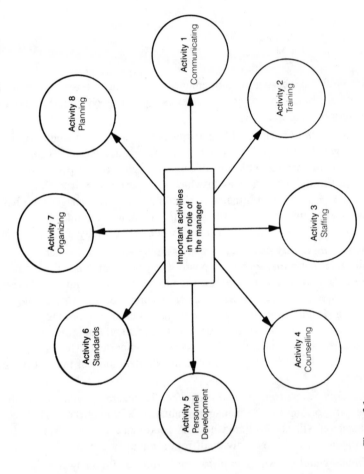

Figure 3.1.
Important activities in the role of manager.

(v) *Humor and emotional control:* These two attributes must also be possessed by a good manager. Good sense of humor and good emotional control play an important role in the daily life of a manager. A short-tempered manager would rarely command the goodwill of persons who work for him.

(vi) *Good listerning ability and self-confidence:* A good manager should also be a good listener rather than talking most of the time when dealing with his or her people. Otherwise, the information coming from subordinates will dry up. Some or all of this information could be very important to the manager. Self-confidence is another characteristic of a good manager. Poor self-confidence of a manager will be reflected in inadequate productivity of his or her department.

(vii) *Ability for self-appraisal and recognizing different points of view:* A manager who possesses the self-evaluation ability has a clear-cut advantage over his or her counterparts because such managers will be able to identify his or her shortcomings and improve upon them. This way his or her performance will be tremendously improved.

Recognition of differing views is another asset a good manager must possess. A manager who is able to recognize that occasionally there may be more than one view which is correct will command a high respect from his or her subordinates.

(viii) *Quickly seeing goodness in people and eagerness to pass credit to others:* A good manager should be able to identify quickly the good points of people working for him or her. Furthermore a good manager should be eager to give credit to his or her subordinates for their successes. It is usually a good practice to meet more than halfway when passing such credits.

3.6 HINTS TO RELIEVE TENSIONS

In the day-to-day work environment, engineers and managers are prone to tensions. There is no mathematical formula which will help to reduce their tensions. However, here we present a number of ways from Reference [6] which will help to reduce tensions common to engineers and managers. Some of these hints are as follows:

(i) Try to behave as a moderate person.

(ii) Try to live each day.

(iii) Try to develop hobbies.

(iv) Try to divide time between work, social life and family intelligently.

(v) Try to periodically reexamine and readjust, if necessary, the set goals.

(vi) Try to be unselfish with respect to others.

(vii) Try not to concentrate on failures by looking into past successes.

3.7   MOTIVATING ENGINEERING MANPOWER

One way to improve the performance of a department is to motivate people working in the department. Results of several studies indicate that a self-motivated employee is far more productive than a poorly motivated employee. In numerical terms these studies show that the productivity of a self-motivated man is two to ten times higher than his counterpart, the poorly motivated man [7]. Therefore motivating employees is also a vital component of a manager's job. This section presents six motivational factors which were developed by Paul J. Meyer [7], who was an insurance salesman and became a millionaire at the age of 27. He opened a success motivation institute in the sixties. The motivators outlined by Meyer are as follows:

(i) Make sure that the employees are involved in the goal setting. If the employees are involved in the goal-setting act then the truly inspired ones will try to achieve their set goals by putting forth their best effort.

(ii) Make sure that the performance measuring standards are set high but within achieving limits of employees. Allow the employees to meet these standards themselves—in other words, without any outside interference.

(iii) Consult employees on matters which are normally above their level. This way one makes them believe that their manager respects their opinions. Furthermore, these people may come up with excellent ideas.

(iv) Make sure that employees clearly understand goals and policies. People who do not understand clearly the objectives or policies will not be that motivated to meet such objectives or policies. Remember that many managers cite communication problems with their subordinates as their most pressing problem.

(v) Make sure that those motivational tools which will be most useful to an individual employee are uncovered. In other words, the same motivational tools may not be able to motivate all the employees because no two individuals are identical in their thinking or have the same objectives. These motivational tools can be uncovered only after knowing the employee well through the "personal touch."

(vi) Make sure that the self-defeating attitudes of certain employees are challenged.

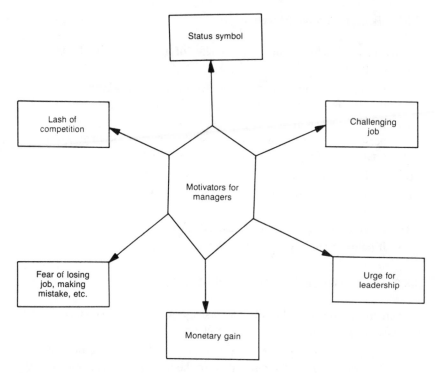

Figure 3.2.
Motivators for managers.

3.7.1 Motivators for Managers

In the previous section we discussed the motivators basically for those people who are in non-managerial ranks. The motivators present in this section are directed at the managers [8]. Motivators are specified in Figure 3.2.

3.8 STAFF MEETINGS

Managers spend a significant portion of their work time in meetings. Furthermore, many important decisions are made in the staff meetings. A competent manager goes to meetings not only to discuss and solve work problems, but also to discover and develop management potential of his or her employees so that at the moment of need the company will not find it difficult to look for management persons.

This section is divided into two parts as shown on next page.

3.8.1   Benefits of a Staff Meeting

This section briefly discusses the benefits of a staff meeting. These are as follows [9]:

(i) It gives a manager an assurance that he does not overlook the potentially gifted executives.

(ii) It is the testing ground to find out the management capabilities of a person. The performance of a person in the staff meetings will give a fair idea of a person's management potential.

(iii) It allows the manager to monitor his or her entire staff in action.

(iv) It helps to identify those persons who are there to find solutions to their problems.

(v) It serves as a barometer of staff members' attitudes.

(vi) It helps to identify the fast thinkers.

3.8.2   Useful Guidelines to Hold Staff Meetings Effectively

There are various guidelines for the manager which help to hold staff meetings effectively. Some of these are as follows:

(i) Encourage participation from all staff members who are present in the meeting. Advance circulation of the meeting agenda helps to increase participation.

(ii) Control the meeting in such a way so that it does not get out of control and the objectives of the meeting are fully accomplished.

(iii) Stimulate the employees' interest in staff meetings so that they do not feel that the meetings are a waste of time.

(iv) Encourage the use of visual aids in the meetings whenever one is presenting facts, figures and so on.

(v) Discourage disruptions when the meeting is in progress.

(vi) Examine the performance of actions and decisions which were decided during the previous meeting. If a manager does not examine the performance of past decisions then the value of staff meetings will be greatly reduced.

(vii) Whenever the manager (chairperson) has the opportunity to assign seats to participants, he should carefully examine the seating position of the potential troublemaker. According to Reference [10] and behavioral researchers, the troublemaker should be seated on position "A" shown in Figure 3.3. The idea behind this arrangement is that the meeting's chairman most probably (according to statistics) is a right-eyed dominant person. This way he or she can tactfully ignore the troublemaker when he or she tries to participate. On the

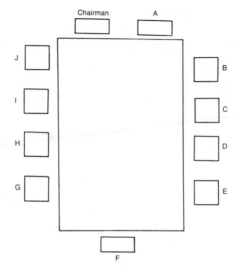

Figure 3.3.
Seating layout for a staff meeting.

other hand, when a person wishes to challenge the chairperson he should sit on seat F specified in Figure 3.3. This way he or she can confront the chairperson with eye-to-eye contact. Furthermore, if the person seated in position F keeps his or her hand up to participate in the discussion then it can be very disconcerting to the meeting chairperson. Therefore the chairperson has to avoid such seating arrangements.

3.9 THE COMMITTEES

In the democratic and communist worlds, committees are widely used to make various decisions. Sometimes a committee is known as "task force," "board," or "commission." A committee is composed of people to whom, as a team, certain matter is committed [8].

In the day-to-day work environment one frequently hears voices against the committees. However, according to the survey of subscribers of the *Harvard Business Review*, which was reported in Reference [11], only 8 percent of the responding subscribers would get rid of committees if they ever had power to do so.

Functions of committees are evaluating policies, recommending and implementing solutions, reviewing performance, generating alternatives, etc.

3.9.1   Why Have Committees?

According to Reference [8], there are various reasons for having committees. Some of them are as follows:

(i) *To reduce power of a single person:* This is one of the main reasons to have common use of committees because of the fear of giving too much authority to a single individual.

(ii) *To share and transmit information:* This is another reason for having committees. With the use of committees, the information can be shared uniformly and faster by the members of the group.

(iii) *To encourage motivation through participation:* Committees are helpful in motivating members of the committee because they participate in the policy-making process. People who take part in the decisions are usually enthusiastic to carry them through.

(iv) *To represent various concerned groups:* Another reason for forming committees is to have participation of those groups which will be concerned with the decision.

(v) *To coordinate activities among several groups of an organization:* In addition, the committee also becomes useful to coordinate planning.

(vi) *To have judgement of a team:* This is probably the most important reason to form committees because the decisions made by a group usually are regarded better than the decisions taken by a single person. The reasons for this belief are a wider range of experience of the group members, a more thorough examination of facts by the group members, and so on.

(vii) *To delay decision on a problem:* Managers or others in some situations appoint committees in order to tactfully delay the decision on a problem. Furthermore, this way, in some cases, the decisions are even postponed indefinitely.

3.9.2   Drawbacks of Committees

Just like anything else, a committee has its disadvantages as well. Some of them are as follows:

(i) *It splits responsibility:* Members of a committee do not feel as much responsibility for the group action as they would have felt if such action was taken by them individually.

(ii) *It consumes a considerable amount of time to arrive at a final group decision.*

(iii) *It is costly in monetary terms.*

(iv) *It is subject to minority tyranny:* Usually the committee decisions are expected to be unanimous or almost unanimous. Because of this the minority members have a strong influence on the final decision of the committee.

3.10 DISPLACING MANAGERS

Sometimes the displacement of a manager from his or her present job is necessary because of poor performance and other factors. Therefore when a person in a managerial position is removed from his or her position, the removal has to be very carefully managed. Otherwise, it may damage the company image in the eyes of its employees, public relations and so on.

There are basically five ways to displace a manager from his or her present job according to Frank L. Bird [12], who served as a manufacturing manager in a big U.S. industrial enterprise. In *Business Topics*, published by Michigan State University, he outlined these five ways, as shown in Figure 3.4:

  (i) Up (North)
 (ii) Down (South)
(iii) Centre
(iv) Out (West)
 (v) Lateral (East)

All of these ways are described below in detail.

UP

This is sometimes known as the move to the north. It is the easiest way in which the manager is removed from his or her present position: giving him or

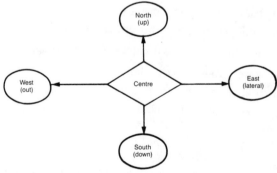

Figure 3.4.
Ways to remove a manager.

her a promotion. The affected individual most probably will not even become aware of the motive behind such a move and may interpret it as a genuine promotion. This type of move is practiced in those situations where the concerned individual has become powerful by having strong connections with the public, customers, etc. If such a move is not successful then it may be necessary to force out the manager from the organization.

Advantages and disadvantages of the upward move are as follows.

Advantages

(i) No damage to the morale of the concerned individual

(ii) No damage to the morale of his or her work team

(iii) Easier to get the approval of the affected individual because of bigger title and probably even higher financial gains

(iv) The move may be disguised to the company, affected individual, etc.

(v) It is the easiest move.

Disadvantages

(i) It is a costly move because the salary of the individual most probably will be increased or maintained at the same level.

(ii) Affected individual may interfere with the functioning of his or her old group because of group members' loyalty toward him or her.

DOWN

This move is known also as a move in a southern direction. In this case the concerned person is demoted from his or her present position. This approach is usually practiced with only those individuals who will take it well or are old or sick. Usually a younger person will not take such humiliation that easily and still be loyal to the organization.

Advantages of this method are:

(i) The company will probably pay a lower salary.

(ii) The person is removed from his or her old position.

However on the other hand, the disadvantages of this approach are:

(i) Possible lower productivity of the individual

(ii) Reduction in attitude of loyalty to the organization

(iii) Possible poor effect on the morale of other members of group or company

CENTRE

In this case the individual's office and title are retained but his or her group is removed. In other words, the individual's group is attached to another department. Usually, it is a difficult move to perform.

This kind of move is usually practiced in a situation where the removal of the manager has to be kept secret from the public.

Disadvantages of this approach are as follows:

(i) The move may affect the morale of the individual's group which is being attached to some other department.

(ii) Productivity of the group members may decrease.

(iii) There may be no reduction in the individual's salary.

OUT

This is known also as a move in a western direction. In this move the individual is forced to leave the organization. This is the simplest way. Usually it is practiced as the last resort if the other moves fail. Sometimes the other four moves are practiced so that the individual leaves the company on his or her will. Furthermore, sometimes the western direction, the "forced out" move, is softened by giving an early retirement or disability leave if situations permit.

Disadvantages of this move are as follows:

(i) Possibility of poor effect on the morale of certain groups of people

(ii) Possibility of adverse publicity

(iii) Possibility of adverse effect on investors and customers because they may assume the existence of unstable forces in the enterprise

LATERAL

In this case the individual is moved laterally in an eastern direction to a position such as that of a staff assistant or a special assistant. The affected person is paid the same salary. When conditions permit, the same individual may be put back into his or her original position. Some of the advantages of the lateral move are as follows:

(i) The affected person is frequently receptive to such a move.

(ii) Chances for the publicity of the move are rather low.

(iii) Execution of such a move is relatively easy.

(iv) The morale and loyalty of the individual to the company may not change if the assigned lateral position will be of real value to the concerned individual.

3.11 SUMMARY

This chapter briefly discusses those aspects of engineering management which are directly or indirectly concerned with the human element. The chapter begins by outlining the needs of an engineer. Sixteen such needs are listed. The next topic discussed is "the routes open to engineer for managerial positions." Seven such routes are described. The next topic which will also interest all engineers is concerned with changes which are expected to occur when an engineer moves to a managerial position. Thus, seven such changes are discussed.

Activities and attributes of a manager are two other topics which are discussed in the chapter. Eight basic activities and sixteen qualities of a manager are outlined.

Two other topics described in the chapter are the "hints to relieve tensions" and "motivating engineering manpower." Both topics are briefly explored. The staff meetings are the next theme of discussion. The benefits of the meetings are discussed and guidelines to conduct effective meetings are specified. The next topic, "the committees," is related to the topic of meetings. The reasons for having committees are explored along with the disadvantages of the committees.

The last topic is concerned with displacing managers. Five different ways to remove managers from their current positions are described in detail. Advantages and disadvantages of most of these ways are given.

3.12   EXERCISES

1. Describe at least ten attributes of an engineering manager.
2. Suppose that you are a vice-president of an engineering company and one of your managers is not performing his or her managerial duties properly. What are the actions you would consider to displace such a person from his or her present position?
3. Discuss at least eleven important needs of an engineer.
4. What are the ways open to an engineer to become a manager?
5. What are the functions of an engineering manager?
6. Discuss the disadvantages of appointing committees.
7. How would you motivate executives?
8. What are the reasons for setting up a "commission" or a "task force?"

3.13   REFERENCES

1. Glueck, W. F. *Personnel: A Diagnostic Approach*. Plano, TX 75075:Business Publications, Inc., pp. 5–6 (1982).
2. Karger, D. W. and R. G. Murdick. *Managing: Engineering and Research*. 200 Madison Ave., New York 10016:Industrial Press, Inc. (1969).
3. Crane, R. K. "What It Is to Be a Boss," in *Management Guide for Engineers and Technical Administrators*. N. P. Chironis, ed. New York:McGraw-Hill Book Company, pp. 5–7 (1969).
4. Couey, F. "What Makes a Good Manager?" in *Management Guide for Engineers and Technical Administrators*. N. P. Chironis, ed. New York:McGraw-Hill Book Company, p. 8 (1969).

5. Lessells, G. A. "Attributes Good Manager Should Possess," in *Management Guide for Engineers and Technical Administrators*. N. P. Chironis, ed. New York:McGraw-Hill Book Company, p. 9 (1969).

6. Klautsch, A. A. "10 Ways to Relieve Tensions Common to Engineers," in *Management Guide for Engineers and Technical Administrators*. N. P. Chironis, ed. New York:McGraw-Hill Book Company, pp. 18-19 (1969).

7. Meyer, P. J. "How to Motivate Key Men," in *Management Guide for Engineers and Technical Administrators*. N. P. Chironis, ed. New York:McGraw-Hill Book Company, pp. 56-57 (1969).

8. Koontz, H., C. O'Donnell and H. Weihrich. *Management*. New York:McGraw-Hill Book Company, pp. 645-646 (1980).

9. Auger, B. Y. "Use Staff Meetings to Discover and Develop Executive Potential," in *Management Guide for Engineers and Technical Administrators*. N. P. Chironis, ed. New York:McGraw-Hill Book Company, pp. 52-53 (1969).

10. Kelly, J. *How Managers Manage*. Englewood Cliffs, NJ:Prentice Hall, Inc. (1980).

11. Tillman, R. "Committees on Trial," *Harvard Business Review*, 38(3):6-12, 162-173 (1960).

12. Bird, F. L., "How to Displace Executives," in *Management Guide for Engineers and Technical Administrators*. N. P. Chironis, ed. New York:McGraw-Hill Book Company, pp. 63-64 (1969).

CHAPTER 4

Creativity

4.1 INTRODUCTION

The development of our civilization to its present stage is the result of creative thinking of men and women of the past and present. Typical examples of the past discoveries and inventions of creative thinking which have shaped our present world are the wheel, the steam engine, the telephone, automobile, nuclear power, radio and television, and the aeroplane. The important discoveries and inventions are generally not accidents, as they may appear to many of us [1], but are the results of intense attention and determination of innovators and researchers, which were overlooked by ordinary minds.

Reference [2] states that some companies estimated that as high as 80 percent of their sales is due to those products which were not in the market ten years ago. Therefore, it shows the importance of creativity in the present-day world. Furthermore, it may be said that for an organization to stay in business effectively, it has to produce innovative products on a continuous basis.

Broadly speaking, creativity is the ability to come up with innovative results from nature [1].

In the present-day world, various techniques are practiced to obtain creative ideas. One of these techniques is widely known as group brainstorming, the name given by Alex Osborn [3]. According to Osborn, this technique is not that new. This type of approach has been practiced by the Hindu religious teachers of India for over 400 years.

So far, a vast amount of literature has appeared on the subject of creativity. Reference [4] presents a list of 6,823 publications for the period 1566–1974. This book [4] has categorized these publications into eight separate classifications. These are creativity of women, general creativity, scientific creativity, creativity and psychopathology, creativity in the fine arts, creativity in industry, engineering and business, developmental studies, and facilitating creativity through education.

This chapter briefly describes the various aspects of creativity.

4.2   EXAMINATION OF SELECTIVE FACTORS IN CREATIVITY

This section briefly examines three pertinent factors which are related to creativity. These factors are as follows [3]:

(i) Education

(ii) Sex

(iii) Age

Osborn in Reference [3] states that the scientific studies for creative aptitude indicate that there is very little or no difference, among persons of the same age, between those with a college education and those without. Furthermore, Osborn quotes L. L. Thurstone that high intelligence of a person is not the same as being gifted in creative thinking. By taking into consideration the past discoveries and inventions, it is added that many important ideas have been due to those persons who were devoid of specialized training in the problem concerned. The invention of the telegraph by Morse is one such example: Morse was a professional portrait painter. Another example is the invention of a new shell-fragment detector. This was the creation of an unscientific person [3].

According to many findings quoted in Reference [3], the female's creative ability is not inferior to that of men. Furthermore, many scientific tests indicate that women on average have faired better in imagination than their counterparts, the men.

According to Osborn [3], as he reasoned by citing several examples in his book, older persons are no less imaginative than younger ones. In fact, as he quotes from W. Somerset Maugham, "Imagination grows by exercise." Furthermore, Lawton quoted in Reference [3] that the creative imagination is ageless. However, on the other hand, Tangerman [1] states that the experts say the best ages for basic innovations among the various specialists are as follows:

(i) Medical and biological scientists: 40 to 45 years

(ii) Mathematicians and theoretical physicists: 30 to 35 years

(iii) Experimental physicists: 35 to 40 years

4.3   CREATIVE PROBLEM-SOLVING STEPS

There are basically six steps associated with the creative problem-solving process [3,5]. These steps are concerned with problem definition, preparation, idea production, idea development, evaluation, and adoption. All six are briefly discussed below:

(i) *Problem Definition:* This is concerned with the identification of the problem.

(ii) *Preparation:* This is concerned with necessary data collection and analysis.

(iii) *Idea Production:* This calls for generation of ideas.

(iv) *Idea Development:* This is concerned with choosing the most appropriate ideas.

(v) *Evaluation:* This is concerned with the verification of tentative solutions.

(vi) *Adoption:* This is concerned with the final solution implementation.

4.4   WAYS TO DEVELOP CREATIVITY

This section presents six ways to develop one's creativity [3]. These are as follows:

(i) *Reading:* The appropriate sort of reading is very healthy for creative thinking. According to Walt Disney, magazines such as *National Geographic, Reader's Digest* and *Holiday* are rich in vitamins.

(ii) *Experience:* This is the richest fuel for creativity. Experience can be divided into two categories, i.e., first-hand experience and second-hand experience. The first type of experience provides a richer fuel for creativity than the second type. The second type of experience is gained through listening, superficial reading and so on.

(iii) *Games:* Playing appropriate games properly can help to develop creativity. We should note here that, according to Reference [3], out of almost 250 kinds of sedentary games only about half of them entail imaginative problems.

(iv) *Writing:* According to Osborn [3], writing also helps to develop creativity. He has argued to prove this by citing several real life examples.

(v) *Fine Arts and Hobbies:* Fine arts and certain hobbies call for imagination. Therefore they are useful in developing creativity.

(vi) *Finding Solutions to Creative Exercises:* This means practicing creativity by solving creative problems. According to Reference [3], persons who have taken courses in creative problem-solving are found to be far better in the production of good ideas than the ones who have not.

4.5   CHARACTERISTICS OF CREATIVE ENGINEERS AND MANAGERS

This section lists the characteristics of both creative managers and highly creative engineers. Many of their characteristics are the same. However, they are listed separately, as follows.

4.5.1   Highly Creative Engineers

According to Reference [6], a highly creative engineer possesses the following characteristics:

(i) Observation power
(ii) Concentration power
(iii) Belongs to either upper or lower decile of engineering classes in the universities and other institutions (as evidenced by some research)
(iv) Faces ambiguous situations more easily
(v) Questions problems and new ideas
(vi) Has fewer close friends
(vii) Displays more self-confidence
(viii) Independent in thought
(ix) High desire for freedom
(x) Less anxious
(xi) Independent in action
(xii) Thinks in more abstract and theoretical terms
(xiii) More stable

4.5.2   Creative Managers

A creative manager normally possesses the following characteristics, by which he can easily be identified [2]:

(i) Persistent
(ii) Creative memory
(iii) Self-motivation
(iv) Analysis and synthesis ability
(v) Tolerance of ambiguity

- (vi)   Concentration power
- (vii) Ability to toy with ideas
- (viii) Displays originality
- (ix)   Self-confidence
- (x)   Sensitive to problems
- (xi)   Actively curious
- (xii) Fearless of failure
- (xiii) Basically optimistic
- (xiv) Openness to feelings
- (xv)   Self-competitive
- (xvi) Free of jealousy
- (xvii) Non-authoritarian
- (xviii) Fearless of authority
- (xix) No resentment toward authority
- (xx)   Respects others' rights
- (xxi) Open and direct when dealing with others

Many more characteristics of managers may be found in Reference [2].

4.6   CLIMATE FOR CREATIVITY

To obtain better creative results from the engineering department, the management has to introduce the right climate into those departments. Reference [6] presents a number of steps which help to ensure an ideal environment for creativity. Some of them are as follows:

- (i)   Give necessary freedom and opportunity to deserving engineers.
- (ii)   Provide necessary facilities to highly creative engineers.
- (iii) Try to send those engineers who are highly creative to professional conferences and seminars. This way they can interact with their outside counterparts.
- (iv)   Avoid resisting new ideas.
- (v)   Conduct informal seminars.
- (vi)   Assign to individual engineers those problems which best suit his or her field of interest.
- (vii) Recognize that the highly creative engineers have to be financially rewarded.
- (viii) Reward highly creative engineers with status.
- (ix)   Recognize differences in personality. For example, a creative individual may have a habit of certain work hours or a need for privacy.

(x) Try to give recognition to new ideas as soon as possible. Keep the originators of the ideas informed, as much as possible, of decisions related to the data.

4.7 ATTRIBUTES OF A MANAGER OF CREATIVE PEOPLE

A manager of a group of creative people has to have certain attributes. Otherwise the group's creativity may not be developed to its full potential. Therefore, this section presents attributes of an ideal manager of creative people [2]. Some of them are as follows:

(i) Possesses knowledge of the creative process

(ii) Gives credit to others when it is necessary

(iii) His subordinates accept him as one of their colleagues rather than as a boss.

(iv) Possesses the necessary professional knowledge to lead his group

(v) Assigns responsibilities to an individual in such a way which is compatible to his or her field of interest

(vi) Able to criticize the work of his people in such a way that their feelings are not hurt

(vii) Gives recognition whenever it is deserved

(viii) Possesses self-confidence

(ix) Provides inspiration at the moment of need

(x) Leads by suggestion

(xi) Is able to take calculated risks whenever it is necessary

(xii) Possesses ability to identify the problem

(xiii) Able to handle people with differing personalities

(xiv) Ability to communicate effectively with others

(xv) Keeps close contacts with the top management by informing it on a regular basis

(xvi) Receptive to all ideas

(xvii) Pushes for flexible organization

4.8 BARRIERS TO CREATIVE THINKING

This section briefly discusses certain barriers which may inhibit creativity in any personality. The effective creativity results can be obtained only once these barriers are removed. Therefore, the management and concerned persons have to be aware of such mental blocks. For practical purposes there are basically four types [6,7] of such barriers as shown in Figure 4.1. All these barriers are briefly discussed below:

(i) *Type I: Environmental Barriers:* These barriers are associated with an individual's development. For example, if an individual was fed with solutions to all his problems by his or her parents, then that individual is apt to be dependent upon other people to find solutions to his or her problems. The same reasoning may be applied to cases such as student-teacher, boss-subordinate, etc.

(ii) *Type II: Cultural Barriers:* These are due to the society in which he or she functions. Therefore, the cultural barriers include all those blocks which are imposed on an individual by the society. For example, in American society, factors such as competitiveness, conformity and cooperation play an important role in shaping one's creative thinking.

(iii) *Type III: Perceptual Barriers:* These are due to the mind's tendency to short-circuit. In other words, these occur because of failure to recognize each situation's components as individual elements. For example, when a person becomes very much acquainted with some item, then he tends to overlook background details of such an item.

(iv) *Type IV: Emotional Barriers:* These are the most serious of all the blocks so far discussed. Examples of such barriers are fear, hate, desire for security, self-satisfaction, greed, over-cooperation, love, resistance to change, poor self-confidence, unwillingness to accept help from others and so on.

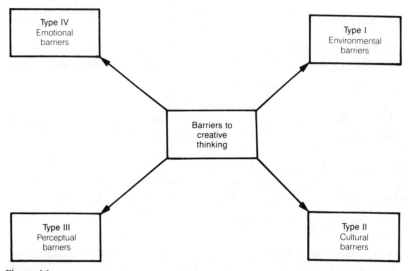

Figure 4.1.
Barriers to creative thinking.

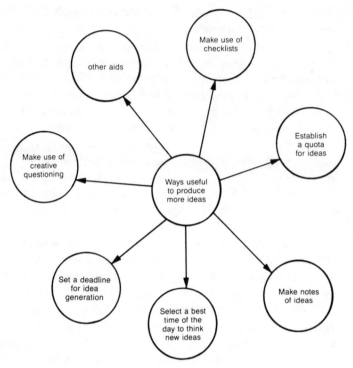

Figure 4.2.
Ways helpful to generating more ideas.

4.9   GENERATION, PRESENTATION AND EVALUATION OF NEW IDEAS

This section separately discusses the ways to generate more ideas, and the presentation and evaluation of ideas. All these are briefly described below.

4.9.1   Generation of New Ideas

To generate new ideas there are no magic formulas which can be used to fulfill such a goal. However, experimentally it is demonstrated that there are procedures which help to increase an individual's new idea production [7]. These are shown in Figure 4.2. The methods of ideas generation shown in this figure are self-explanatory, and therefore, they are not described here. However, the detailed description of each of them is given in Reference [7].

4.9.2   Presentation of New Ideas

This is one of the most important aspects of the creative process. If the presentation of even a brilliant idea is not effective then it may die. Therefore, in order to increase the idea acceptance probability, the presentation of an idea requires thorough planning and preparation. Reference [2] suggests idea presentation guidelines such as sell the new idea first to your boss, obtain information on the audience, sell by implanting, gain information on conditions that facilitate idea selling, and look for proper timing when selling a new idea. In the actual presentation of new ideas, most of the factors one should take into consideration are as follows:

   (i) Show muted enthusiasm.

   (ii) Be prepared for objections.

 (iii) Present the historical aspect of the problem and the procedure of your solution.

 (iv) Do not follow the argumentative procedure.

  (v) Stay in good control and calm.

 (vi) Try to be as natural as possible.

 (vii) Give more attention to the practical aspects of your idea.

(viii) Welcome any suggestions which will help to improve the idea.

 (ix) Communicate your idea in such a way that the language used is concise, clear and simple; in other words, so that it can easily be understood by your audience.

  (x) Add a light touch of showmanship.

 (xi) Avoid claims which are unsubstantiated.

 (xii) Leave out a vital detail of the idea purposely hoping for others to fill it in. This way they will feel they have contributed to the idea as well. If no one else fills in the missing gap then try to fill it in by yourself, but at the proper time.

(xiii) When your audience includes technically or professionally sophisticated persons, then try to introduce counterarguments to your idea as well and discuss them.

(xiv) Neutralize objection to the new idea as follows:

    (a) Ask such a question to the person who has objected to the idea, with which he or she has to agree.

    (b) Listen to the person who has objected to the idea carefully and encourage him to elaborate on the objection. Generally, it is found that the more a person talks, the weaker his argument against the idea becomes.

 (xv) Outline the immediate steps appropriate for the new idea implementation.

(xvi) Try to arrange for a break after the presentation. It will give time to the audience to sort out questions related to the presentation. After the break, try to give all answers to questions arising from the audience. At the end, sum up the presentation and advantages of the idea and the reason for its implementation.

4.9.3   Evaluation of New Ideas

Each new idea has to go through an evaluation process before its implementation. Therefore, it is advisable to the new idea thinkers to evaluate it themselves before wasting much time and money. According to Reference [2], which quotes L. M. Moore of the Massachusetts Institute of Technology, each new idea can be classified into one of the following three categories:

   (i) *Difficult ideas:* These are those ideas which can be accomplished only in twelve months or more time and require a great deal of other efforts.

  (ii) *Moderate ideas:* These ideas can probably be completed in six months with a reasonable sum of investment.

 (iii) *Simple Ideas:* These ideas do not require an extensive effort or expense and can be accomplished immediately.

According to Reference [2], one chemical company uses factors such as marketability, growth, engineering, production, research and development, stability and company's position to evaluate new product lines. After taking all of these factors into consideration, each new product idea is rated into one of five categories. These are poor, very poor, average, good and very good.

4.10   WAYS TO KILL IDEAS

A manager plays an important role in his department's creativity effort. If the manager is receptive to new ideas of his subordinates then the creativity effort of his group will be effective. Otherwise, he may be killing good ideas which may be worth millions of dollars to his organization one day. Here are some of the negative ways a manager may act towards his subordinate who comes up with a new idea:

   (i) It will be difficult to sell.

  (ii) It is a ridiculous idea.

 (iii) It will be very costly.

  (iv) The top management will reject it.

   (v) It will not be beneficial to us because it is not our problem.

  (vi) Your idea is too radical.

 (vii) It is outside our responsibility.

(viii) We are not big enough to try this idea.

  (ix) We do not have time for such ideas.

   (x) It is too theoretical.

  (xi) We are doing quite well without it.

 (xii) Our present budget would be unable to accommodate it.

(xiii) We have tried similar ideas before.

(xiv) Nobody else has tried it before.

Most of the above negative responses were selected from Reference [8]. In this reference, the responses were obtained from a brainstorming session conducted at the New York chapter of the American Society of Training Directors. All in all, a good manager has to avoid such negative responses to obtain healthy creativity results from his group.

4.11   CREATIVITY TECHNIQUES

This section presents various techniques to generate new ideas and solve creativity problems. These are as follows.

4.11.1   Technique I

This technique is known as the Gordon method, after its originator W. J. Gordon [9]. With the aid of this technique, the Gordon team was able to develop items such as a new type of razor, a toothbrush, a gasoline pump and a can opener. This approach has two characteristics [10]:

  (i) The team explores the underlying concept of the problem. For example, if a new can opener is desired then the team will explore the subject of opening.

 (ii) The team attacks the underlying concepts at length. Furthermore, the subjects are investigated from various aspects.

In this approach the team would first discuss, for example, the meanings of the word opening and examples of opening in real life things. This way unusual approaches will be revealed. At the later stage, the team attacks and develops the uncovered approaches.

The advantages of this approach are as follows:

  (i) The radical applications of existing methods are encouraged.

 (ii) The early closure on the problem is prevented.

Gordon's team, which developed various items, was composed of six people [5]. Generally, the technique session requires, on average, one day of discussion.

4.11.2   Technique II

This is known as the CNB method [5]. The assumptions made in this technique are that the team consists of competent persons who understand the objective of the problem and are willing to cooperate. This technique is composed of the following steps:

(i) Distribute to each member of the team a notebook containing a problem of major scope, preparative material, and creative aids.

(ii) Allow each member of the team to record his thoughts on how to solve the problem for a month. During each day of the month the team member writes one or more of his or her ideas on solving the assigned problem. At the end of the month each member selects the best idea and writes fruitful suggestions for further exploration in that notebook. In addition, remaining new ideas are also summarized in the notebook.

(iii) Each team member submits his or her notebook to the coordinator at the specified time.

(iv) The coordinator studies all the notebooks and prepares a detailed summary.

(v) After the summarization, any member of the team can see all the notebooks.

(vi) Whenever it is necessary, a final meeting is held in which all team members participate.

This technique is described in detail in Reference [5].

4.11.3   Technique III

This is known as the "attribute-listing" approach and is due to Professor R. P. Crawford [3]. In this approach a person is required to list attributes of an idea or an item. Once the attribute list is completed then the person concentrates on one attribute at a time, in order to make improvements on items or ideas to which these attributes belong. For further information on this method consult Reference [3].

4.11.4   Technique IV

This is known as the "forced relationships" method and is due to Charles S. Whiting [3]. The main objective of this approach is to generate new ideas by creating a forced relationship between two or more usually unrelated ideas or items. That is how the idea-generation process starts. In Reference [3], this approach is demonstrated by giving the example of an office equipment manufacturer whose list of manufactured products includes desk lamp, bookcase, desk, chair and so on. Thus, in this case the idea thinker can consider the relationship between the chair and desk as a starting point. From then on, the idea thinker would seek to start a line of free associations. Through this approach he may end up with ideas for a new product, for example, a combined unit composed of both desk and chair. At the end, the most promising idea is selected.

4.11.5   Technique V

This is known as the brainstorming method and is due to Alex Osborn [3], who applied it on the modern line in 1938. However, Osborn says that this technique is not that new. This type of approach has been practiced by the Hindu religious leaders for over 400 years. These teachers called it Prai-Barshana. The meaning of the term Prai is "outside yourself" and the meaning of the term Barshana is "question." However, the meanings of the modern term "brainstorming" is "using the brain to storm a problem." A group of people participate in the brainstorming sessions. In these sessions one idea for solving a problem triggers another idea and the process continues. The time spent in each session is always less than an hour. Sometimes it is as short as 15 minutes. Persons who participate in these sessions have different backgrounds but similar interests.

According to Reference [11], the best results are achieved when 8 to 12 persons participate in each session. In addition, it is recommended that one should aim to achieve at least 50 ideas in each session. The following guidelines are to be observed when conducting a brainstorming session:

(i) *Welcome Free-Wheeling:* The wilder the idea is, the better it is. It is easier to discard ideas than to think them up.

(ii) *Aim for Quantity:* It is better to have a large number of ideas than a small number of them because the probability of finding useful ideas out of a large number is greater than from a small number.

(iii) *Eliminate Idea Criticism During the Session:* Because the unwanted ideas will be eliminated during the screening process, make sure that criticism of each other's ideas does not take place during the session.

(iv) *Combine and Improve Ideas:* At the end of the session ask the participants how the ideas obtained during the session can be turned into better ideas by combining and improving.

(v) *Record Ideas:* During the brainstorming session the ideas must be recorded; otherwise they will be forgotten.

(vi) *Keep the Ranks of Participants Fairly Equal:* For example, a supervisor will need a lot of warming up time to mix his or her ideas with those of his directors.

(vii) *Choose the Time of the Session Carefully:* Hold the sessions in the morning only if the participants are going to work on the same problem in the remaining work hours as well. Otherwise, conduct these sessions in the afternoon. This way the participants' regular work productivity will not be reduced.

An example of a topic for a brainstorming session is, "What are the ways to kill new ideas?" According to Reference [8], one session generated 56 different ways.

4.11.6   Technique VI

This is another brainstorming method and is known as the tear-down approach [5,11]. The technique can be used by two persons. After picking the subject for brainstorming, person A takes the attitude that the existing way is incorrect and then contributes an idea suggesting an alternative way, but not necessarily better.

The second person, person B, is not allowed to agree with the idea of person A. Therefore, person B has to suggest another way. Similarly, person A is forbidden to agree with person B and has to suggest another way. This cycle continues until a useful idea clicks.

4.11.7   Technique VII

This is another brainstorming technique and is known as the and-also method [5,11]. Again, this approach can be used by two persons. The approach is the same as technique VI, but with one exception: that both persons must agree with each other and then make an addition to the other's idea. For example, after selecting a topic for the brainstorming, person A presents an idea to make improvement on the subject under study. Person B agrees with the idea of person A and improves on A's idea. Similarly, person A agrees with the idea of person B and improves on person B's idea. The cycle continues until a sound solution is reached.

4.11.8   Technique VIII

This technique can be used by a single person [11]. In this approach the problem is written and distributed to various participants before the meeting. In addition, it is specified that each participant has to have a certain number of solution ideas, say 17, to the problem before he is allowed to attend the meeting. This way a large number of tentative solution ideas may be obtained. Thus a sound solution to the problem may be found.

4.12   SUMMARY

This chapter briefly presents the various aspects of creativity. The pertinent factors related to creativity are examined. These factors are education, sex and age. Creative problem-solving steps are outlined. In addition, ways to develop creativity are briefly described. Creative people have specific characteristics. Therefore, the chapter lists the characteristics of creative engineers and managers.

The creativity can flourish in a group only if the group is functioning under ideal environments. Therefore the chapter presents a number of steps which help to ensure ideal climate for creativity. Two other topics discussed in the chapter are the attributes of a manager of creative people and the barriers to creative thinking. The chapter lists 16 attributes of the manager of creative people and describes four stumbling blocks to creativity.

The important topics of new idea generation, presentation and evaluation are discussed.

The creativity of a group depends also upon how a manager reacts to new ideas of his subordinates. The undesirable actions of a manager may kill good ideas. Therefore, the chapter presents a list of undesirable responses of a manager which will kill the new ideas. Lastly, eight creativity techniques are described. The techniques include the Gordon method, CNB method, brainstorming, attribute-listing, forced relationships and so on.

The source references for the material presented in the chapter are listed.

4.13 EXERCISES

1. What are the factors which help to develop one's creativity? Describe them in detail.
2. Describe briefly the creative problem-solving process.
3. Write an essay on the historical aspect of creativity.
4. List the characteristics of the creative person.
5. What are the factors which help to develop an ideal climate for creativity of a group?
6. What are the stumbling blocks to creativity?
7. What are the factors that you would take into consideration when making the presentation of new ideas?
8. How would you go about evaluating new ideas?
9. What are the meanings of the term brainstorming? Describe the brainstorming approach.
10. Compare the brainstorming method with Gordon's technique.
11. Describe the following creativity techniques:
    (i) The and-also method
    (ii) The tear-down approach
    (iii) The attribute-listing procedure

4.14 REFERENCES

1. Tangerman, E. J. "Creativity: The Facts Behind the Fad," in *Management Guide for Engineers and Technical Administrators*. N. P. Chironis, ed. New York:McGraw-Hill Book Company, p. 262–265 (1969).

2. Randsepp, E. *What the Executive Should Know About Creating and Selling Ideas*. Larchmont, NY:American Research Council (1966).

3. Osborn, A. F. *Applied Imagination*. New York:Charles Scribner and Sons (1963).

4. Rothenberg, A. and B. Greenberg. *The Index of Scientific Writings on Creativity*. Hamden, CT 06514:Archon Books, The Shoe String Press (1976).

5. Haefele, J. W. *Creativity and Innovation*. New York:Reinhold Publishing Corporation, pp. 12–13 (1962).

6. Karger, D. W. and R. G. Murdick. *Managing: Engineering and Research*. 200 Madison Ave., New York 10016:Industrial Press, Inc., pp. 91–92 (1969).

7. Mason, J. G. *How to Be a More Creative Executive*. New York:McGraw-Hill Book Company, Inc., pp. 47–67 (1960).

8. "How to Kill Ideas," in *Management Guide for Engineers and Technical Administrators*. N. P. Chironis, ed. New York:McGraw-Hill Book Company, Inc., p. 227 (1969).

9. Gordon, W. J. *Synectics*. New York:Harper & Brothers (1961).

10. Von Fange, E. K. *Professional Creativity*. Englewood Cliffs, NJ:Prentice-Hall, Inc. (1959).

11. Studt, A. C. "How to Set up Brainstorming Sessions," in *Management Guide for Engineers and Technical Administrators*. N. P. Chironis, ed. New York:McGraw-Hill Book Company, pp. 276–277 (1969).

CHAPTER 5

Manpower Planning and Control

5.1 INTRODUCTION

Manpower planning and control is one of the essential activities of management for the effective functioning of an organization, whether it is an engineering or non-engineering organization. The poor management of manpower planning and control activity may be reflected in the quality of output of a company, morale of the manpower employed, efficiency of the organization and so on. In today's environments, functioning effectively is far more demanding for an organization than in the past because of stiff competition for the availability of engineering manpower, size of the organization, complexity of the product produced, cost and so on. Therefore, we may say that during the past decade or so the manpower planning and control functions have achieved maturity. Nowadays, management practitioners are making use of quantitative techniques to make manpower planning and control decisions because, in recent times, many advances have been made in the mathematical aspect of manpower planning and control. Therefore, this chapter is devoted to the subject of manpower planning and control.

The main objective of this chapter is to present selective books and journals in which the manpower planning and control literature may be found and to make the engineering management practitioners or students aware of the existence of manpower planning and control mathematical models. Therefore, the chapter presents a classification of published literature on manpower planning and control, basics of manpower planning and control, and selective mathematical models concerning the topic under discussion. The emphasis of the chapter is on mathematical models.

5.2 CLASSIFICATION OF PUBLISHED LITERATURE ON MANPOWER PLANNING AND CONTROL

A large number of publications have appeared on the topic of manpower planning and control in many forms, e.g., books, journal publications and conference papers. These publications cover a wide spectrum of activities associated with manpower planning and control. Therefore, this section is

65

concerned with the classifications of such published literature. Most of the published literature may be grouped into nine categories. These are as follows:

 (i) *Span of Control:* This basically is concerned with the supervision of manpower by one individual. Several articles may be found in the published literature on this topic alone.

 (ii) *Organizational Size and Efficiency:* This is another aspect of manpower planning and control concerned with the size of the organization and its efficiency. This topic is discussed in several publications in the open literature.

 (iii) *Labor stability:* This is concerned with the manpower turnover. A number of articles may be found in the published literature on this topic.

 (iv) *Manpower Planning and Forecasting:* This is concerned with strictly planning and forecasting manpower need. A large number of articles are written on the topic.

 (v) *Manpower Selection and Recruitment:* This deals with selection and recruitment of manpower. The topic is covered in various publications.

 (vi) *Probabilistic and Stochastic Models:* These mathematical models are concerned with manpower planning. There are a significant number of journal articles presenting various kinds of stochastic models.

 (vii) *Case Studies:* These report the result of real life studies conducted in various parts of the world and within specific industries. A number of papers may be found on manpower planning and control.

 (viii) *Review Articles:* These articles review various aspects of manpower planning and control. A number of such articles may be found in the published literature.

 (ix) *Miscellaneous Publications:* These publications do not fall into any of the above groups but are concerned with various aspects of manpower planning and control. A number of published articles fall into this classification.

Books and journals in which the above categorized literature may be found are listed [1–33] at the end of the chapter.

5.3   MANPOWER PLANNING AND CONTROL

The subject of manpower planning and control has been receiving increased attention in recent years because of considerable growth in cost of the wages and salaries and due to many other factors. The objective of manpower planning may be defined in various ways; however, according to Reference [34], the overall objective may be stated as bringing an organization's manpower

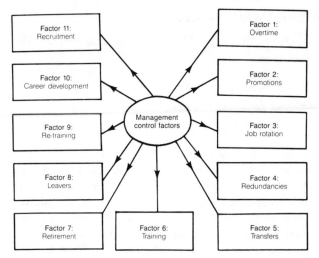

Figure 5.1.
Management control factors.

into line with the present requirement of the company and its need for a period ahead. To fulfill this objective, plans for some of the eleven management control factors given in Figure 5.1 will definitely be needed.

5.3.1   Approaches for Manpower Planning

There is no single manpower planning approach which may be suitable for all situations. Furthermore, various people have presented different approaches or procedures to tackle the problem of manpower planning. Therefore, this section describes briefly two procedures used for manpower planning, which are essentially based on the similar lines.

MANPOWER PLANNING PROCEDURE I

This approach, taken from Reference [34], is essentially divided into six steps. These steps are as follows:

(i) Review manpower operations
(ii) Identify corporate strategy
(iii) Forecast the manpower demand
(iv) Forecast the manpower supply
(v) Reconcile demand and supply forecasts
(vi) Exercise control

The above six steps are self-explanatory, and therefore no discussion is presented. However more discussion on these points may be found in Reference [34].

MANPOWER PLANNING PROCEDURE II

This is another procedure used for manpower planning [35]. This approach has only three phases:

(i)   Review of present manpower and study of external factors
(ii) Manpower resources in future
(iii) Monitoring progress results

This procedure states that manpower planning is a continuing operation and therefore the above three phases must be carried out on a continuous basis.

The study of present manpower resources is concerned with the study of labour turnover, organizational structure, recruitment and selection, age distribution, skills, personal records, absenteeism, training, performance evaluation and so on. In addition, external factors such as national economic situation, organizational change, duty to community and so on are studied during this phase.

Future manpower resources deal with forecasting vacancies, forecasting future manpower (redundancies, labor losses, etc.), impact of mechanisation and automation in future, career planning, labor relationship and so on.

The third and final phase of this approach is concerned with monitoring progress, which includes manpower reviews, audits and so on.

5.4   SELECTIVE MATHEMATICAL MODELS

This section presents mathematical models which will directly or indirectly aid in making useful manpower planning and control decisions.

Therefore, the mathematical models associated with span of control, labour stability, organization size and efficiency, and learning process are presented. The reasons for the selection of these models were simplicity, ease in collecting data, practicality and so on. In this chapter, the aim of the author is to present the end resulting equations of these models rather than develop them from the first principle. However, the source references are given which present the full derivations of those models in question. These models are as follows.

5.4.1   Model I – Span of Control

This model [36] can be used to calculate the number of persons, p, to be supervised by a leader; the total number of leaders, L_t, in an organization; hierarchy levels, k, in an organization; and the work force, F_w, of an organization (excluding supervisory personnel). The total number of leaders in a company is given by

$$L_t = \frac{F_w(p^k - 1)}{p^k(p - 1)} \tag{5.1}$$

If the company has one president, then the above equation reduces to

$$L_t = (p^k - 1)/(p - 1) \tag{5.2}$$

because

$$\frac{F_w}{p^k} = 1 \tag{5.3}$$

Rearranging Equation (5.3) leads to the following equations:

$$k = (\log F_w)/(\log p) \tag{5.4}$$

and

$$p = \sqrt[k]{F_w} \tag{5.5}$$

The following points are to be noted for this model:

(i) Span of control is the same at all levels.
(ii) The organizational levels decrease as the span of control increases.
(iii) From Equation (5.4) it can be concluded that k is inversely proportional to $\log p$ because the F_w is a constant.

The results obtained from this model will provide input into making span of control decisions.

EXAMPLE 5.1

An engineering organization employs 4,000 workers at the shop floor level. The number of organizational levels above the working level is equal to 20. Assume that the company has one president and the same number of people are supervised by each person at all levels. Calculate the number of persons to be supervised by a leader.

In the example the given values of F_w and k are as follows:

$$F_w = 4,000$$

$$k = 8$$

Thus utilizing the data in Equation (5.5) results in

$$p = (F_w)^{1/k} = (4000)^{1/8} = 2.83 \cong 3 \text{ persons}$$

Each leader should supervise three persons.

5.4.2 Model II – Labour Stability

This is simply an index which is used to assess the stability of the work force [37,38] in an organization. The stability index, S, is defined as follows [37]:

$$S = T_s/0.5 \, m \, t \tag{5.6}$$

where

m denotes the number of persons employed by the organization at present.
T_s denotes the presently employed persons' total length of service in years.
t denotes the time in years between the mean retirement age of employees and the mean recruitment age.

If the value of this index is less than unity then it indicates that the company work force is not sufficiently stable. However, on the other hand, if its value is above unity then it clearly shows that the work force is excessively stable. It means that when a large percentage of employees reach the retirement age together, the company may be faced with hiring and retention problems. More information on this and other related indices may be found in Reference [37].

EXAMPLE 5.2

An electronic components manufacturer employs 400 persons. The total length of service of all persons employed by the company is 6,000 years. The time between the mean retirement age and the mean recruitment age is 25 years. Calculate the value of the stability index.

In this example the values of T_s, t and m are defined as follows:

T_s = 6,000 years
t = 25 years
m = 400 employees

By substituting the above given data for T_s, t and m in Equation (5.6) we get:

$$S = \frac{6,000}{(0.5)(400)(25)} = 1.2$$

The value of the stability index, S, is 1.2. Therefore, it indicates that the company manpower is excessibly stable.

5.4.3 Model III – Organization Size and Efficiency

This mathematical model [39] is concerned with determining the relationships between organization size and efficiency. The model is directed primarily towards organizations which basically deal with paper study oriented tasks such as research. Examples of such organizations, including Johns Hopkins Operations Research Office and the Rand Corporation, are given in Reference [39].

The total number of publications or reports, k, produced by a department or an organization annually is given by

$$k = (240 \; \beta)/(T_w + T_r \; \beta \; \mu) \qquad (5.7)$$

where

β represents the total number of professional persons employed in the organization or the department.

μ denotes the fraction of all publications (reports) received by the average professional person; in other words, those reports which the person in question is expected to read.

T_r denotes the mean time (days) to read one report by a professional person.

T_w denotes the mean time (days) to accomplish one report by a professional employee. This time includes the time spent on investigation, analysis, writing and so on.

It is assumed that in one year there are 240 workdays. As the number of professional persons, β, becomes very large, the value of k in Equation (5.7) approaches the upper limit; i.e.:

$$k' = 240/\mu \; T_r \qquad (5.8)$$

The efficiency, E, of the organization is defined by

$$E = \frac{k}{k_0} = [1 + (\mu \; \beta \; T_r)/T_w]^{-1} \qquad (5.9)$$

where k_0 denotes the number of reports which can be produced if no time was spent for reading any report.

For very large β, the value of E tends toward zero value, E being inversely proportional to the value of β. Further developments on the model may be found in Reference [39].

EXAMPLE 5.3

A consulting organization's basic task is to conduct various types of engineering research. The company employs 500 professional persons to carry out such tasks. Each employee works eight hours per day and 240 days per year. In addition, each professional employee spends on average 2 days to read a report written by others and 25 days to write his report. The writing time includes time spent on investigation, analysis, writing, and so on. Each professional worker reads only 1/5 of the total reports received per year. Calculate the total number of reports to be produced annually by the company in question.

In this example the values of β, μ, T_w and T_r are as follows:

β = 500 professional employees
μ = 1/5
T_w = 25 days
T_r = 2 days

Utilizing the above data in Equation (5.7) leads to

$$k = \frac{240\ (500)}{25 + (2)\ (500)\ (0.2)} = 533.33 \text{ reports/year.}$$

Therefore, the company will produce approximately 533 reports annually.

5.4.4 Model IV—The Learning Curve

According to Reference [40], the learning curve concept seems to have originated in the aircraft industry. It was observed in the aircraft industry that as the output of a specific type of airplane was increased, the per unit average direct labour input decreased on a regular basis.

Therefore this model is based on the fact that the more frequently a person or a worker repeats a specified task, the more efficient that person will become [7,40]. In this case the time reduction results from the learning phenomena. In this time reduction process, factors other than the learning factor may also contribute; they are job design, better tooling, better equipment, improved design and so on. The following equation is used to represent the time reduction curve.

$$Z = t_f\, y^{-\alpha} \tag{5.10}$$

where

Z denotes the cumulative average man-hours per item.
y denotes the quantity of items produced.
t_f denotes the time taken to produce the first item.
α represents the curve exponent.

In the aircraft industry it was found that if an aircraft's first unit took 2000 man-hours to manufacture, the second unit took only 1600 man-hours (in other words, to produce two units, the time will be 3200 hours instead of 4,000 hours), the fourth unit absorbed 1280 man-hours, the 8th unit required 1024 man-hours, the 16th unit took 819.2 man-hours and so on. It means that to double production of the airplane units it required only 80% of the previous time. For example, to produce the fourth unit, it took only 80% of man-hours (i.e., 1280 man-hours) of second unit (i.e., 1600 hours). Therefore in the aircraft industry the 80% learning factor is widely practiced.

To estimate the value of α in Equation (5.10) for the 80% learning factor we assume that the time to produce the first unit and the second unit is given by the following equations, respectively:

$$Z_1 = t_f \, y_1^{-\alpha} \tag{5.11}$$

and

$$Z_2 = t_f \, y_2^{-\alpha} \tag{5.12}$$

where

Z_1 is the time to produce first unit, y_1.
Z_2 is the time (per unit) to produce second unit, y_2.

Dividing Equation (5.11) by Equation (5.12) results in:

$$\frac{Z_1}{Z_2} = \frac{y_2^{\alpha}}{y_1^{\alpha}} \tag{5.13}$$

To double production at 80% learning factor we let $y_2 = 2y_1$, $Z_1 = 2,000$ man-hours, and $Z_2 = 1600$ man-hours in Equation (5.13) as follows:

$$\frac{2,000}{1600} = \frac{(2y_1)^{\alpha}}{y_1}$$

$$\frac{5}{4} = 2^{\alpha} \tag{5.14}$$

Thus, taking logarithms of Equation (5.14) and rearranging leads to

$$\alpha = 0.3219$$

Thus, at 80% learning factor, Equation (5.10) is rewritten in the following form:

$$Z = t_f \, y^{-0.3219} \tag{5.15}$$

Similarly, the calculated values of α at 95%, 90%, 85% and 75% learning factors are 0.07, 0.15, 0.23, and 0.42, respectively. Thus one may rewrite Equation (5.10) for these factors, respectively, as follows:

For 95% Learning Factor:

$$Z = t_f \, y^{-0.07} \tag{5.16}$$

For 90% Learning Factor:

$$Z = t_f \, y^{-0.15} \tag{5.17}$$

For 85% Learning Factor:

$$Z = t_f \, y^{-0.23} \tag{5.18}$$

For 75% Learning Factor:

$$Z = t_f \, y^{-0.42} \tag{5.19}$$

The plots of Equation (5.10) for the various given values of α are shown in Figure 5.2. These curves show that the per unit (or task) time decreases as more units (or tasks) of the same type are produced (or performed). The decrease is rapid for the first ten units (or tasks). More information and the procedure for estimating the value of α are given in Reference [7].

EXAMPLE 5.4

A company is to produce 16 identical parts of an engineering system. To manufacture the first part requires 2,000 man-hours. When the production of the units is doubled, the per unit manufacturing time is reduced by 80% (i.e., the 80% of the time before doubling the production).

Calculate the total number of man-hours needed to manufacture 16 parts. Utilizing the given data in Equation (5.15) yields:

$$Z = t_f \, y^{-0.3219} = (2,000) \, (16)^{-0.3219}$$

$$= 819.2 \text{ man-hours}$$

This means to produce 16 units, the per unit average manufacturing time will be 819.2 man-hours. Thus to manufacture 16 components, the total time, T, is equal to

$$T = (819.2) \, (16)$$
$$= 13,107.2 \text{ man-hours}$$

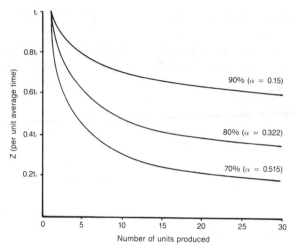

Figure 5.2.
Time reduction curves.

5.5 SUMMARY

This chapter covers various aspects of manpower planning and control. The published literature on manpower planning and control is classified into nine separate categories. These are span of control, organizational size and efficiency, labor stability, manpower planning and forecasting, probabilistic and stochastic models, manpower selection and recruitment, case studies, review articles and miscellaneous publications. Over 30 sources of literature on manpower planning and control are listed at the end of the chapter. Eleven management control factors are presented along with two manpower planning procedures. In addition, the chapter presents four mathematical models selected from the published literature. These models are concerned with span of control, labor stability, organizational size and efficiency and the learning process. All four models are supplemented with numerical examples along with their solutions.

5.6 EXERCISES

1. Discuss the objective of manpower planning.
2. What are the management control factors? Describe them briefly.
3. Discuss the main commonalities and differences between the manpower planning procedures I and II presented in the chapter.
4. A small engineering company employs 500 workers at the shop floor level. The organization has one president and it desires that six persons are to be supervised by each supervisor in the company.

Calculate the number of organizational levels needed above the shop floor level (or the working level).

5. The total length of service of all employees in an engineering consulting organization is 2500 years. The company employs a total of 250 persons. The mean recruitment age of all employees is 40 years and the mean retirement age is 62 years. Calculate the value of the stability index and comment on its value.

6. An engineering research organization employs a very large number of professional employees. The research organization produces 1000 reports per year. The value of μ in Equation (5.8) is equal to 1/10. Calculate the value of mean time (in days) each employee spends reading a report.

7. Prove that the value of α in Equation (5.10) for the 65% learning factor is equal to 0.622.

5.7   REFERENCES

1. Bartholomew, D. J. and A. F. Forbes. *Statistical Techniques for Manpower Planning.* New York:John Wiley & Sons (1979).

2. Bartholomew, D. J. and B. R. Morris. *Aspects of Manpower Planning.* New York:American Elsevier Pub. Co. (1971).

3. Burack, E. H. and J. W. Walker. *Manpower Planning and Programming.* Boston:Allyn and Bacon Inc. (1972).

4. Clough, D. J., C. C. Lewis and A. L. Oliver, eds. *Manpower Planning Models.* London:English University Press (1974).

5. Grinold, R. C. and K. T. Marshall. *Manpower Planning Models.* New York:North-Holland (1977).

6. Holt, C. C., F. Modigliani, J. F. Muth, H. A. Simon and C. P. Bonini. *Planning Production, Inventories and Work Force.* New Jersey:Prentice-Hall (1960).

7. Howell, J. E. and D. Teichroew. *Mathematical Analysis for Business Decisions.* Homewood, IL:Richard D. Irwin, Inc. (1971).

8. Kruisinga, H. J. *The Balance Between Centralization and Decentralization in Managerial Control.* H. E. Stenfert Kroesen, V.-Leiden (1954).

9. Machol, R. E., A. Charnes, W. W. Cooper and R. J. Niechaus, eds. *Management Science Approaches to Manpower Planning and Organizational Design.* Amsterdam:North-Holland (1978).

10. Simon, H. A. *Models of Man: Social and Rational.* New York:John Wiley (1957).

11. Vadja, S. *Mathematics of Manpower Planning.* New York:John Wiley (1978).

12. *British Journal of Industrial Relations,* Houghton St., Aldwich, London EC2A 2AE, England:London School of Economics and Political Science.

13. *Business Horizons,* School of Business, Indiana University, Bloomington, USA.

14. Association of European Operational Research Societies. *European Journal of Operational Research Society.* Box 211, 1000AE Amsterdam, Netherlands:North-Holland Publishers.

15. *Harvard Business Review.* Soldiers Field, Boston, MA 02163:Graduate School of Business Administration, Harvard University.

16. *IEEE Transactions on Engineering Management.* New York:Institute of Radio Engineers.

17. *Industrial Relations Journal*, Waterloo Road, London SE1 8UL, England:Mercury House Publications Ltd.
18. *International Journal of Manpower*, 198/200 Kingley Road, Bradford, BD9 4JQ, England:Institute of Personnel Management, MCB Publications.
19. *IRE Transactions on Engineering Management*, New York:Institute of Radio Engineers.
20. *Journal of Human Resources*, University of Wisconsin Press, 114 N. Murray St., Madison, WI 53715:Industrial Relations Research Institute.
21. *Journal of Industrial Economics*, 108 Cowley Road, Oxford OX4 1JF, England:Basil Blackwell Publishers Ltd.
22. *Journal of Management Studies*, School of Administration, University of Ghana, BOS 78, Achimota, Legon, Ghana.
23. *Journal of Operational Research Society*, Maxwell House, Fairview Park, NY 10523, USA:Pergamon Press, Inc.
24. *Management Science.* Institute of Management Science, 146 Westminister St., Providence, RI 02903, USA.
25. *Omega.* Maxwell House, Fairview Park, NY 10523, USA:Pergamon Press Inc..
26. *Organizational Behaviour and Human Performance.* 111 Fifth Avenue, New York, NY 10003, USA: Academic Press Inc..
27. *Operational Research Quarterly*, Maxwell House, Fairview Park, NY 10523, USA:Pergamon Press Inc..
28. *Personnel*, American Management Associations, AMACOM Division, 135 W. 50th St., NY 10020, USA.
29. *Personnel Management*, Waterloo Road, London SE1 8U1, England:Mercury House Publications Ltd.
30. *Personnel Psychology*, 198/200 Kingley Road, Bradford BD9 4JQ, England:MCB Publications Ltd.
31. *Personnel Review*, Institute of Personnel Management, MCB Publications, 198/200 Kingley Road, Bradford BD9 4JQ, England.
32. *Review of Economics and Statistics*, Harvard University USA, Box 211, 1000 AE, Amsterdam, Netherlands:North-Holland Publishers.
33. *The American Economic Review*, American Economic Association, Suite 809, Oxford House, 1313 21st Ave. S., TN 37212, USA.
34. Sawtell, R. A. and P. Sweeting. "A Practical Guide to Company Manpower Planning," *Personnel Review*, 4(4):33–40 (1975).
35. Rowe, K. *Management Techniques for Civil Engineering Construction.* New York:John Wiley & Sons (1975).
36. Meij, J. L. "Some Fundamental Principles of a General Theory of Management," *The Journal of Industrial Economics* (1):16–32 (Oct. 1955).
37. Bowey, A. "A Measure of Labour Suitability," *Personnel Management*, 3(4):26–32 (April 1971).
38. Bowey, A. M. "Labour Stability Curves and a Labour Stability Index," *British Journal of Industrial Relations*, 7(1):71–83 (Mar. 1969).
39. Adler, F. P. "Relationships Between Organization Size and Efficiency," *Management Science*, 7(1):80–84 (Oct. 1960).
40. Hartley, K. "The Learning Curve and Its Application to the Aircraft Industry," *Journal of Industrial Economics*, 2:122–128 (Mar. 1965).

CHAPTER 6

Selecting Engineering Projects

6.1 INTRODUCTION

Selection of projects is very important to engineering organizations, because through the evaluation process the decisions concerning investment in engineering projects can be made intelligently. Many of the companies in business depend on such decisions. In the engineering industry each day, many companies are faced with the selection of the most desirable project out of many alternative projects. The project selection process is not an easy task because it is surrounded by various uncertainties. Therefore, the project selection process makes use of the application of various useful procedures and models to reduce the unforeseen risks. According to Reference [1] the objectives of project selection are four-fold, as follows:

(i) Reject the unsuitable engineering projects
(ii) Select only those projects which are viable and will bring in satisfactory return on the investment
(iii) Make sure that the desirable projects are not rejected unless it is necessary
(iv) Make sure that there is no damaging effect on flow of project proposals

To achieve the above objectives requires the installation of appropriate procedures and good judgements from management. Otherwise the engineering companies will have a damaging effect on their business.

This chapter covers the various aspects of selecting engineering projects.

6.2 PROJECT SELECTION FACTORS

This section briefly presents some of the factors which influence the selection of many research and development projects. The three main categories of these factors are shown in Figure 6.1.

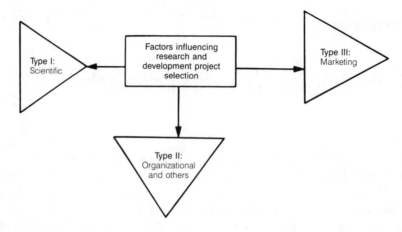

Figure 6.1.
Main categories of factors influencing research and development project selection.

Figure 6.1 categories are described below [2]:

TYPE I

This category contains the scientific related factors such as:

 (i) Project success probability
 (ii) Competition
 (iii) Research discoveries' patentability
 (iv) Utilization of existing research manpower
 (v) Time period involved in achieving the project end goal
 (vi) Compatibility with other on-going research programs
 (vii) Existing resources' utilization
 (viii) Cost involved in completing the entire project
 (ix) Possibility of gaining technical expertise foothold in that area
 (x) Impact on other program plans
 (xi) Usefulness in future activities

TYPE II

The factors listed under this category are concerned with organizational and other aspects. These are as follows:

 (i) Company's prestige
 (ii) Moral duty
 (iii) Project timing
 (iv) Capabilities in manufacturing
 (v) Pressures from outside bodies

(vi) Impact on organizational attitudes
(vii) If the project is terminated, the capability to absorb scientific manpower
(viii) In the event of termination of the project under consideration, the room to absorb the concerned project facilities

TYPE III

These factors are concerned with the marketing aspect and are listed below:

(i) Expected profit
(ii) Compatibility with the marketing activities
(iii) Forecasted sales for the production in question
(iv) Expected profit from the product
(v) Consumers' satisfaction with the current competitive products in the market
(vi) Influence from those products which are currently being developed

6.3   PROCEDURES FOR ENGINEERING PROJECT SELECTION AND FEASIBILITY ANALYSIS

This section presents two procedures for selecting desirable engineering projects for development. The first procedure is used to perform the project feasibility analysis, whereas the second procedure is essentially a mathematical method used to select the right new product idea.

6.3.1   Procedure I

The feasibility analyses are performed to make rational decisions regarding the investment in new projects. The analyses are concerned with probing the feasibility of the project under consideration. Furthermore, the ultimate objective of these analyses is to see if the venture is going to be a profitable one. The formality of the feasibility analysis is dictated by the size and type of the project. Generally the feasibility anlaysis serves to give information on items such as:

(i) Adequacy of labor supply
(ii) Availability of raw materials
(iii) Need of the product in the market
(iv) Technical soundness of the project
(v) Financial viability

According to References [3,4] it may be said that the project feasibility anal-

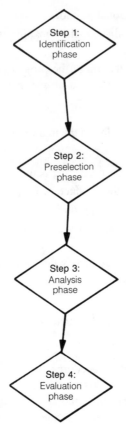

Figure 6.2.
Phases of project feasibility analysis.

ysis is essentially composed of four basic steps as shown in Figure 6.2. All these steps or phases are described in detail below.

IDENTIFICATION PHASE

This is the first step in the project feasibility analysis and is concerned with clarifying the objective to be accomplished. For example, the objective may be to find out if it is possible and profitable to develop a specific kind of product.

Some of the sources for the ideas and needs for new engineering product development are as follows:

(i) Salespersons

(ii) Market studies

(iii) Studying the new legislation effects

(iv) Survey of existing industry

(v) Competitors

(vi) Distributors

(vii) New technology implication study

(viii) Published literature

PRESELECTION PHASE

Broadly speaking, the objectives of this phase are as follows:

(i) To find out whether it is worthwhile to perform project feasibility analysis in detail. If yes, then

(ii) To outline scope of the studies to follow

(iii) To find out how much it will cost for the subsequent studies

Thus the following two items are accomplished during this phase:

Preliminary Screening of Ideas: It will not be possible to perform prefeasibility analysis on each and every idea proposed for new venture. Therefore the preliminary screening is conducted to eliminate ideas which are expected to be unsuccessful at the later stage. The preliminary screening is composed of two steps as follows:

(i) Go/no-go procedure for idea elimination

(ii) Comparative rating analysis

The go/no-go procedure for idea elimination is based on answering seven questions given in Figure 6.3. If the answer is yes to any one of these questions for a particular idea, then that idea should be eliminated.

Only those ideas which first pass through the go/no-go elimination procedure successfully go through the comparative anlaysis procedure. The areas covered in the comparative analysis are risks, market, costs and market potential. Thus these topics are evaluated in detail subjectively in the comparative analysis. A checklist of items to be considered in risks, market, costs and market growth potential is specified in Reference [3].

Prefeasibility Analysis: Once the preliminary screening is accomplished, the attractive ideas obtained from the result of this study go through the prefeasibility analysis. This analysis is conducted to further screen the ideas which have passed through the preliminary screening. Furthermore, the prefeasibility analysis is conducted because the complete feasibility analyses are costly and time-consuming to perform. Generally, the following are the main components of the prefeasibility anlaysis:

(i) Profit and cost estimates

(ii) Product and market descriptions

(iii) Technological variants description

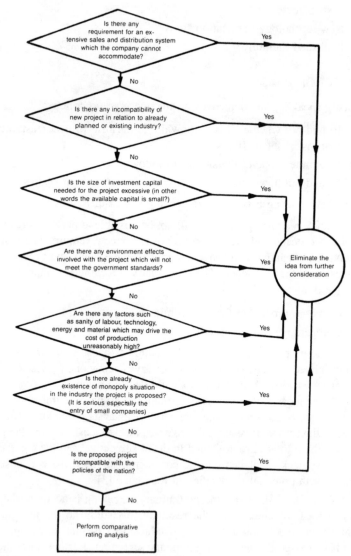

Figure 6.3.
Seven test questions for eliminating proposed ideas.

(iv) Identification of plant location factors
(v)   Availability of pertinent production needs such as skilled manpower, energy, materials and water
(vi) Capital requirement estimate
(vii) Identification of major risks and problems
(viii) Miscellaneous data

During the course of the prefeasibility study, when the information warrants that the new idea being investigated is to be rejected, the prefeasibility analysis should be discontinued. The data used in the prefeasibility analysis is usually collected on an informal basis through interviews, trade associations and published literature.

ANALYSIS PHASE

This is the third step of the feasibility analysis and is concerned with the analysis aspect. The analyses conducted in this phase may be divided into three main groups known as the market, technical and financial analyses.

The market analysis is basically concerned with evaluating a project's success in relation to the market. In comparison to the marketing analysis used for screening, the market analysis conducted during evaluation is more thorough. The objectives of the technical analysis are three-fold:

(i)   To investigate the technical feasibility of the project under consideration
(ii) To study the effect of technical alternatives on factors such as balance of payments, employment of manpower, etc.
(iii) To provide a base to estimate costs

Information on materials, manpower availability, product, capital availability and market has to be available before the technical analysis can begin.

The third component of the analysis phase is concerned with the financial aspect; for example, preparing financial statements basically to determine cost and profitability of the project in question.

EVALUATION PHASE

This is the final step of the project feasibility analysis. Usually during this phase social profitability analyses and investment proposals for funding are prepared. The quantitative social profitability analysis must be performed if there are national priorities. If the results of the social profitability study are favorable, then the investment proposal for funding is prepared. Otherwise the project is terminated.

6.3.2   Procedure II

This is a mathematical technique to select a desirable project for development. The technique is composed of seven steps as shown in Figure 6.4. The technique begins by assuming all ideas for new projects are on an equal basis. All the ideas arriving at each step are evaluated with the aid of an established acceptance decision criterion of the company in question. If an idea passes through the decision process successfully, then the idea is sent to the next stage of the selection method. Otherwise it is filed for future reference. The main objective of this storing process is that after a period of time the stored idea may look like an attractive venture for development.

The advantage of this approach is that it reduces risk associated with the new product development ideas. But the disadvantage is that the successful products have to pay for all the drop-out products due to this procedure, in addition to their own development expenditure.

DECISION CRITERION USED TO EVALUATE EACH STEP OF THE PROCEDURE

This decision criterion is based on dollars and cents. For example, if a company spends e amount of money (dollars) to produce a new product and in return the income is r dollars due to that product, then the product project will be accepted for development only if the value of $(r - e)$ is positive and if it also meets other requirements specified in the company's decision criterion. According to Reference [5] some of the decision criterion factors which always appeal to investors are as follows:

(i) *Time to break even:* This is concerned with finding out how much time is needed to break even financially with the proposed new product. To use this factor the company must have a set time limit to break even.

(ii) *Gain to risk factor:* This is an important evaluation factor which is utilized at each stage of the procedure under consideration. Generally, the minimum value of the ratio of the gain to risk factor is about 2.5:1.

(iii) *Dollar volume per year:* In this respect the company must set a limit for minimum acceptable dollar volume so that the estimated dollar volume of the new product can be compared.

DESCRIPTION OF PROCEDURE STEPS

This section briefly describes the steps outlined in Figure 6.4. The symbols used in this figure are e_i and r_i for $i = 1,2,3 \ldots ,7$. These symbols (i.e., e_i and r_i) denote the expenditure and income, respectively, in the ith step. The total expenditure of a product is equal to the sum of e_is (i.e., $\sum_{i=1}^{7} e_i$). Similarly, the total income from the same product is given by $\sum_{i=1}^{7} r_i$. Thus the

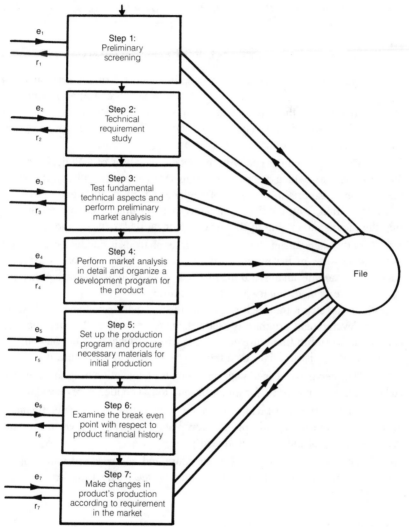

Figure 6.4.
Stages of the product selection procedure.

net profit, P_n, is given by

$$P_n = \sum_{i=1}^{7} r_i - \sum_{i=1}^{7} e_i \qquad (6.1)$$

The steps in Figure 6.4 steps are briefly described as follows:

Step 1: This is a preliminary screening step in which informal market analysis and a brief literature search are conducted. Furthermore, es and rs are estimated. Therefore on this basis the decision is made whether to further explore the idea or simply reject it. The rejected idea is filed. The filed idea may become attractive after some period of time.

Step 2: This step is concerned with the detailed analysis of technical requirements of that idea which has passed through Step 1 successfully. In this stage the opinion of experts in the areas of engineering, production and marketing and information from published literature play an important role. From this step the idea is either sent to the next step or is filed away for the future.

Step 3: In this step a brief, but formal, market analysis is conducted and basic technical features of the product are tested. In addition, es and rs are estimated. After all this analysis, if the product idea still looks promising then the next step is executed; otherwise, it is filed away.

Step 4: In this step a thorough market analysis is performed and a full-scale product development program is established. At the end of this step the idea is examined again to find out whether it is still worthwhile. If the outcome of this examination is negative, then the idea is filed away for future use.

Step 5: This step is concerned with setting up the production program and procuring necessary materials for initial production. The idea is reviewed again.

Step 6: In this step the break-even point with respect to product financial history is examined. In addition, the acceptance criterion is used to examine the profitability of the newly produced product.

Step 7: This is the last step of the procedure and is concerned with making necessary changes in production as dictated by the product market.

6.4 PROJECT SELECTION MODELS

There are various methods, models, indices and formulas developed to select engineering projects [6,7,8,9] for development. Some of them are described below.

MODEL I

This is a widely known model used to determine the maximum amount of justified money for an engineering project. This model was originally developed by Disman [10]. Thus, according to Disman the income, I, over a specific time period from an investment is given by

$$I = \alpha \cdot i + \alpha \tag{6.2}$$

where

i is the rate of return on the investment for the project.
α is the investment.

Thus for one year, rearranging Equation (6.2) in terms of α results in:

$$\alpha = \frac{I}{(1 + i)} \tag{6.3}$$

The above expression is utilized to estimate the value of maximum expenditure justified (E_{mj}) for an engineering project when the annual estimated net income, I_n, and the rate of return are known. Thus, for one year

$$E_{mj} = \frac{I_n}{(1 + i)} \tag{6.4}$$

Similarly, from Equation (6.4), for k years

$$E_{mj} = \sum_{j=1}^{k} \frac{I_{nj}}{(1 + i)^j} \tag{6.5}$$

where I_{nj} is the net income from the project for year j.

To take into consideration the risks associated with the products' technical and commercial successes, Equation (6.5) is modified to the following form:

$$E_{mj} = F_t \cdot F_c \left(\sum_{j=1}^{k} \frac{I_{nj}}{(1 + i)^j} \right) \tag{6.6}$$

where

F_c is the risk of product commercial success.
F_t is the risk of product technical success.

EXAMPLE 6.1

The estimated net incomes from a new product are $50,000 and $40,000 for first and second year, respectively. The value of i before taxes should be 30%. Calculate the maximum expenditure justified for the new product.

Thus substituting the given data into Equation (6.5) yields

$$E_{mj} = \frac{I_{n1}}{(1 + i)} + \frac{I_{n2}}{(1 + i)^2}$$

$$= \frac{50,000}{(1 + 0.3)} + \frac{40,000}{(1 + 0.3)^2}$$

$$= 38,461.5 + 23,668.6$$

$$= \$62,130.10$$

EXAMPLE 6.2

The estimated annual net income from a new engineering product for four years is $60,000. The product technical success risk probabilities are 0.7 and 0.6, respectively. The value of i before taxes should be 30%. Calculate the value of the maximum expenditure justified for the new product.

For $k = 4$, the Equation (6.6) simplifies to

$$E_{mj} = F_t \cdot F_c \left\{ \frac{I_{n1}}{(1 + i)} + \frac{I_{n2}}{(1 + i)^2} + \frac{I_{n3}}{(1 + i)^3} + \frac{I_{n4}}{(1 + i)^4} \right\} \quad (6.7)$$

Thus substituting the specified data into Equation (6.7) results in:

$$E_{mj} = (0.7)(0.6) \left\{ \frac{60,000}{(1 + 0.3)} + \frac{60,000}{(1 + 0.3)^2} + \frac{60,000}{(1 + 0.3)^3} + \frac{60,000}{(1 + 0.3)^4} \right\}$$

$$= (0.42) \{46,153.85 + 35,502.96 + 27,309.97 + 21,007.67\}$$

$$= \$54,589.268$$

MODEL II

This model is used to calculate the payback time over which the project capital expenditure will be recovered. According to Reference [11], this type of approach is often convenient, especially when making decisions regarding the expenditures on design of equipment for processing plants. If the estimated useful life of the product under consideration is equal or little more than the payback time, then the investment for such a project will be unattractive.

However, if the project useful life is much more than the payback time, then the project would certainly be more appealing. The principal disadvantage of the payback time models is that these models do not take into consideration the profits once the capital expenditure has been recovered. This model (II) consists of four different formulas to calculate payback time. These formulas are presented below.

FORMULA A

The payback time, T_{pb}, is defined by:

$$T_{pb} = \frac{\alpha}{P_{ag}}$$ (6.8)

where

α is the capital expenditure.
P_{ag} is the mean annual gross profit.

EXAMPLE 6.3

A company uses a manual machine to produce a certain type of components. An automatic machine tool to replace the manual machine costs $70,000. However, it can save $7,000 per year to produce specified components. The machine is maintenance-free. Calculate capital expenditure recovery time based on mean annual gross profit.

Thus substituting the specified data into Equation (6.8),

$$T_{pb} = \frac{\alpha}{P_{ag}} = \frac{70,000}{7,000} = 10 \text{ years}$$

FORMULA B

This is another formula which is used to calculate project payback time. The formula equation is

$$T_{pb} = \frac{\alpha}{P_{an}}$$ (6.9)

where P_{an} is the mean annual net profit.

The net profit per year is given by

$$P_{an} = P_{ag} - [r_t(P_{ag} - r_d \, \alpha) + i \, (\alpha + \alpha_w)]$$ (6.10)

where

i is the interest rate on the borrowed money.
α_w is the working capital.
r_t is the rate of income tax.
r_d is the invested capital (fixed) depreciation rate.

EXAMPLE 6.4

Assume that in Example 6.3 the following additional data are given:

(i)   Interest rate on the borrowed money is 5% per year.
(ii) The income tax rate after depreciation is 60%.
(iii) The invested capital (fixed) depreciation rate is 15%.

Calculate the capital investment payback time based on mean annual net profit.

In this example the additional data are specified for the following items:

$$i = 5\%, \, r_t = 60\%, \, r_d = 15\%.$$

By utilizing the specified data in Equation (6.10), we get:

$$P_{an} = 7,000 - \{0.6[7,000 - 0.15(70,000)] + 0.05(70,000 + 0)\}$$

$$= 7,000 - [4200 - 6300 + 3500]$$

$$= \$5600$$

Thus, from Equation (6.9)

$$T_{pb} = \frac{\alpha}{P_{an}} = \frac{70,000}{5,600} = 12.5 \text{ years}$$

FORMULA C

This is the third type of formula used to compute payback time of capital investment. The formula is expressed as

$$T_{pb} = \alpha/c_{af} \tag{6.11}$$

where

α is the capital expenditure.
c_{af} is the mean annual cash flow.

Thus,

$$c_{af} = P_{an} + \alpha \, r_d \tag{6.12}$$

EXAMPLE 6.5

In Examples 6.3 and 6.4 the input data for Equations (6.11)–(6.12) are specified. With the aid of such data, determine the value of payback time based on mean annual cash flow.

From Examples 6.3 and 6.4 we get

$$\alpha = \$70,000; \ P_{an} = \$5,600; \ r_d = 15\%$$

Thus the mean annual cash flow is

$$C_{af} = \$5,600 + (70,000) \cdot (0.15)$$

$$= \$16,100$$

Finally, the payback time based on mean annual cash flow is

$$T_{pb} = 70,000/16,100$$

$$= 4.3478 \text{ years}$$

FORMULA D

This formula is concerned with calculating the maximum payback time based on θ, the minimum acceptable rate of return on investment. This payback time is fixed by the company. Thus, the formula is expressed as [11]:

$$T_{mpb} = \{r_d + [\theta/(1 - r_t)]\}^{-1} \tag{6.13}$$

where T_{mpb} is the maximum acceptable payback time.

EXAMPLE 6.6

To purchase electric equipment requires $60,000. The estimated values of depreciation and income tax rates are 15% and 30%, respectively. Furthermore, the maximum acceptable value of rate of return is 20%. Calculate the maximum acceptable payback time.

In this example the data are specified for the following items:

$$r_d = 15\%, \ r_t = 30\%, \ \theta = 20\%.$$

Thus substituting the above data into Equation (6.13) results in

$$T_{mpb} = \{0.15 + [0.2/(1 - 0.3)]\}^{-1}$$

$$= 2.295 \text{ years}$$

MODEL III

This is known as the profitability index and is based on constant demand [12]. The index, I, is expressed as follows:

$$I = \left\{ (F_c F_t) \prod_{i=1}^{3} z_i \right\} (E)^{-1} \tag{6.14}$$

where

z_1 is the mean sales volume on yearly basis.
z_2 is the selling price of a unit.
z_3 is the static market life.
F_t is the technical success probability of the product.
F_c is the commercial success probability of the product.
E is the total expenditure on the project.

MODEL IV

This model is known as the calculated-risk method [13]. The index, I, is expressed as

$$I = \frac{I_{an} P}{E} \tag{6.15}$$

where

I_{an} is the all net incomes per year from the project.
P is the risk factor (which is a composite of total risks associated with incomes).
E is the total expenditure. (This includes capital assets, product and market development.)

This index is the simplified version of return on investment. Therefore, for screening purposes, it becomes quite useful for comparing with a predetermined minimum rate.

MODEL V

This mathematical model is used to calculate return on investment. The model consists of two different formulas. However, both formulas relate the project's net profit to the total expenditure.

In comparison to the payback time approach, the return on investment is complex. The two widely used formulas [13] of this approach are as follows.

FORMULA A

This formula is concerned with calculating the return on original investment. Thus the return on original investment is defined by

$$\beta = \left(\frac{P_{an}}{E + c_w} \right) \times (100) \tag{6.16}$$

where

P_{an} is the mean net profit per year.
E is the amount of money invested.
c_w is the working capital.
β is the return on original investment.

FORMULA B

This is another formula which is used to calculate return on mean investment. Thus, the return on mean investment is expressed as

$$\beta_m = \left(\frac{P_{an}}{\frac{1}{2}E + c_w} \right) \times (100) \tag{6.17}$$

where β_m is the return on mean investment.

EXAMPLE 6.7

An engineering company is considering purchasing a system which will cost $500,000 in investment. In addition, there is a need for an additional $50,000 for working capital. The estimated value of mean net profit per year is $100,000. Determine the return on original investment.

Substituting the specified data into Equation (6.16) results in

$$\beta = \left(\frac{100,000}{500,000 + 50,000} \right) (100)$$

$$= 18.18\%$$

EXAMPLE 6.8

With the aid of already specified data in Example 6.7, calculate the value of return on mean investment.

Thus utilizing Equation (6.17) we get

$$\beta_m = \left\{ \frac{100,000}{(0.5)(500,000) + 50,000} \right\} (100)$$

$$= 33.33\%$$

MODEL VI

This is another model, known as the Manley model, used when evaluating new projects [12]. The basis for the model is the net profit-to-sales ratio. This ratio is expressed as follows:

$$R_{ps} = \frac{(\alpha_1 + \alpha_2 + 0.5\,\alpha_3)}{\alpha_4 \cdot \alpha_5} \tag{6.18}$$

where

α_3 is the cost of research prior to taxes.
α_2 is the expenditure in plant.
α_1 is the working capital.
α_4 is the total sales of the product per year.
α_5 is the time in years to recover total investment.

MODEL VII

This model is concerned with determining the present value of net income from a project. Thus, the present value, W_p, of net income is given [14] by

$$W_p = i + \frac{A}{(I + 1)^{k-1}} \tag{6.19}$$

$$A \equiv I_n/I \tag{6.20}$$

where

i is the interest rate.
k is the time in years to recover capital.
I_n is the net income.
I is the mean value of net return on investment.

MODEL VIII

This model covers the widely known benefit/cost analysis. This analysis is performed to find out if benefits from the project outweigh its cost. More explicitly, it is worthwhile to consider a project for development only if its benefits are greater than the investment cost. Thus the benefit/cost ratio, R_{bc}, is given by

$$R_{bc} = \alpha_b/\alpha_c \tag{6.21}$$

where

α_b is the user benefits.
α_c is the total investment cost (investment cost plus operating cost).

Thus, in this situation the value of R_{bc} must be equal to or greater than unity.

EXAMPLE 6.9

An engineering equipment procurement price is $60,000 and its useful life period is 8 years. The equipment maintenance cost is estimated to $2000 per year. The equipment total benefits in dollars over its entire useful life period are estimated to be $120,000. Determine the value of the benefit/cost ratio.
Thus,

$$\alpha_c = 60,000 + (2,000)(8) = \$76,000$$

and

$$\alpha_b = \$120,000$$

Therefore,

$$R_{bc} = \frac{120,000}{76,000} = 1.58$$

The value of the benefit/cost ratio is 1.58.

MODEL IX

This model is known as the Dean-Sengupta method [15]. Dean-Sengupta index, I_{ds}, is defined as follows:

$$I_{ds} = r_t \cdot F_t \cdot F_c \cdot B_1/B_2 \cdot I_a \qquad (6.22)$$

$$B_1 \equiv \sum_{k=1}^{m} (1 + j)^{-k} \qquad (6.23)$$

$$B_2 \equiv \sum_{k=0}^{m} (1 + j)^{-k} \qquad (6.24)$$

where

j is the rate of interest per year.
m is the project's useful life in years.
r_t is the return from project per year (assuming that the project is a technical success and marketed).
I_a is the investment cost per year.
F_t is the technical success probability of the project.
F_c is the product probability that it will be marketed, if it is successful technically.

Finally, it is added that this index makes it possible to obtain information on the relative performance of two projects by comparing their indices obtained by utilizing Equation (6.22).

MODEL X

This mathematical model is known as the index of relative worth and is defined [12] by

$$I_{rw} = \{F_t F_c + F_e X\}/C \tag{6.25}$$

$$X \equiv \left\{ \sum_{j=0}^{m} (1 + i)^{-j} \cdot y_j \right\} - \left\{ \sum_{j=0}^{m} C_j (1 + i)^{-j} \right\} \tag{6.26}$$

where

m is the product's market life in years.
i is the rate of interest.
C_j is the jth year's investment.
Y_j is the jth year's net income.
C is the investment for research and development of product.
F_t is the technical success probability of the product.
F_c is the product commercial success probability because of new design.
F_e is the product success probability because of economic situations.

MODEL XI

This model is due to Mottley and Newton [16]. Their model is an index which depends on five basic factors. These factors are the strategic need, product completion time, project cost, gain of the market and promise of suc-

Table 6.1 Scores for projects A, B, and C.

Factors	Project A	Project B	Project C
Project Cost (θ_1)	2	1	3
Project completion time (θ_2)	3	1	1
Market gain (θ_3)	1	2	1
Strategic need (θ_4)	3	3	2
Promise of success (θ_5)	3	3	2

cess. According to these authors, the five factors will contribute to the potential profits of the firm. The overall value of the index, I_{mn}, is given by

$$I_{mn} = \prod_{i=1}^{5} \theta_i \qquad (6.27)$$

where

θ_i is the rated score of the ith factor; $i = 1$ (project cost), $i = 2$ (project completion time), $i = 3$ (market gain), $i = 4$ (strategic need), $i = 5$ (promise of success).

Each of the above five factors is given a score from 1 to 3 and then their scores are multiplied to obtain an overall score for the project. Reasons for each factor rating from 1 to 3 are tabulated in References [12] and [16].

EXAMPLE 6.10

A company has three engineering projects, A, B and C, but can select only one for development because of limited resources. For each project, five important factors such as project cost, project completion time, market gain, strategic need and promise of success are specified. After some investigation, their respective factors are rated, as shown in Table 6.1, for all three projects. Determine which is the most promising project to be selected for development.

Substituting the data specified for project A in Table 6.1 into Equation (6.27) yields

$$I_{mn} = \theta_1 \cdot \theta_2 \cdot \theta_3 \cdot \theta_4 \cdot \theta_5 = (2)(3)(1)(3)(3) = 54$$

Similarly, for projects B and C, respectively, the overall scores are as follows:

$$I_{mn} = \theta_1 \cdot \theta_2 \cdot \theta_3 \cdot \theta_4 \cdot \theta_5 = (1)\,(1)\,(2)\,(3)\,(3) = 18$$

and

$$I_{mn} = \theta_1 \cdot \theta_2 \cdot \theta_3 \cdot \theta_4 \cdot \theta_5 = (3)\,(1)\,(1)\,(2)\,(2) = 12$$

Thus, project A is to be selected for development because it has the highest overall score of 54.

MODEL XII

This is due to Sobelman [17]. The model takes into consideration the time value of the money. This is done by finding the present worth of mean annual profits and costs. The equation of the model is expressed as follows:

$$v_p = \alpha_1 A_1 - \alpha_2 A_2 \tag{6.28}$$

$$A_1 \equiv \sum_{k=1}^{m_1} (1 + i)^{-k} \tag{6.29}$$

$$A_2 \equiv \sum_{k=1}^{s_1} (1 + i)^{-k} \tag{6.30}$$

$$m_1 \equiv m + m_2 \left(1 - \frac{s}{s_2} \right) \tag{6.31}$$

$$s_1 \equiv s + s_2 \left(1 - \frac{m}{m_2} \right) \tag{6.32}$$

where

i is the rate of interest.

α_1 is the mean annual net profit.

α_2 is the mean annual cost of the product development.

v_p is the value of project.

m is the product useful life (years).

s is the product development completion time in years.

m_2 is the group of those products' mean useful life period (in years) of which the project is a member.

s_2 is the group of those products' mean development time (in years) of which the project is a member.

We should note in this model that a project must be considered for development only if its value of index is positive.

EXAMPLE 6.11

For the elements of Equations (6.28)–(6.32) the following data are specified:

$$i = 0.1, \ \alpha_1 = \$25,000, \ m = 10 \text{ years}, \ \alpha_2 = \$4,000, \ s = 4 \text{ years},$$

$$m_2 = 5 \text{ years}, \ s_2 = 2 \text{ years}.$$

Determine the value of v_p.

By substituting the specified data into Equations (6.31)–(6.32) we get:

$$m_1 = 10 + 5\left(1 - \frac{4}{2} \right) = 5$$

and

$$s_1 = 4 + 2\left(1 - \frac{10}{5} \right) = 2$$

Thus substituting the above results and the other specified data into Equations (6.29)–(6.30) yields

$$A_1 = \sum_{k=1}^{5} (1 + 0.1)^{-k}$$

$$= \frac{1}{(1.1)} + \frac{1}{(1.1)^2} + \frac{1}{(1.1)^3} + \frac{1}{(1.1)^4} + \frac{1}{(1.1)^5}$$

$$= 3.7908$$

and

$$A_2 = \sum_{k=1}^{2} (1 + 0.1)^{-k}$$

$$= \frac{1}{(1.1)} + \frac{1}{(1.1)^2}$$

$$= 1.736$$

Finally, substituting the above results and the other given data into Equation (6.28) results in:

$$v_p = 25{,}000\ (3.7908) - 4{,}000\ (1.736)$$

$$= \$87{,}826$$

6.5 SUMMARY

This chapter briefly explains the subject of project selection. The chapter begins by describing the four objectives of project selection and then goes on to explain the project selection factors. A list of three types of factors influencing research and development project selection is presented.

The next topics discussed in the chapter are project selection and feasibility analysis procedures. Two such procedures are described: one for the feasibility analysis and the other one for the project selection.

The major emphasis of the chapter is on project selection models. Therefore the entire last section of the chapter is devoted to this topic. A total of 12 models is presented. These models are as follows:

   (i) Disman model
  (ii) Payback time determination model (including four different formulas)
 (iii) Profitability index
  (iv) Calculated-risk method
   (v) Return on investment model (including two different formulas)
  (vi) Manley model
 (vii) Net income present value evaluation model
(viii) Benefit/cost analysis model
  (ix) Dean-Sengupta model
   (x) Index of relative worth model

(xi) Mottley-Newton model

(xii) Sobelman model.

The chapter contains eleven examples along with their solutions. The relevant literature associated with the material presented in the chapter is listed in the reference section.

6.6 EXERCISES

1. What are the important reasons for the project feasibility analysis?
2. Describe the term "prefeasibility analysis."
3. What is meant by the term "working capital"?
4. What are the important factors which influence the selection of a research and development project?
5. What is the difference between the payback time determination model and the return on investment model?
6. An engineering company wishes to develop a machine to produce some engineering components. The estimated yearly net income from that machine is $40,000 for a seven-year period. The rate of return, before taxes, on the investment should be 25%. The machine technical success and commercial success risk probabilities are 0.8 and 0.5, respectively. Determine the maximum expenditure justified for the new machine.
7. An electronic company is investigating a system whose estimated cost would be $1 million. However, the estimated average annual net profit from that machine is $200,000. After some investigation, it was felt that there will be a need for an additional $75,000 for working capital. Determine the values of return on original and mean investment separately.
8. Assume that for the Sobelman's model the following data are specified:

   $s = 5$ years, $s_2 = 2$ years, $m = 13$ years, $m_2 = 6$ years, $i = 5\%$, $\alpha_1 = \$37,000$, $\alpha_2 = \$7,000$.

   Determine the value of the project. Will this project be a worthwhile venture?

6.7 REFERENCES

1. Wearne, S. H. *Control of Engineering Projects*. London:Edward Arnold (1974).
2. Faust, R. E. "Project Selection in the Pharmaceutical Industry," *Research Management*, pp. 46–55 (September 1971).

3. Clifton, D. S. and D. E. Fyffe. *Project Feasibility Analysis*. New York:John Wiley and Sons (1977).

4. "The Stages of Preparation and Implementation of Industrial Projects," *Industrialization and Productivity*, Bulletin of United Nations Industrial Development Organization, No. 19, New York (1973).

5. Huetten, C. and L. Sweany. "A Mathematical Technique for Choosing the Right New-Product Idea," in *Management Guide for Engineers and Technical Administrators*. N. P. Chironis, ed. New York:McGraw-Hill Book Company (1969).

6. Clarke, T. E. "Decision-Making in Technologically Based Organizations, A Literature Survey of Present Practice," *IEEE Transactions on Engineering Management*, EM-21: 9–23 (Feb., 1974).

7. Baker, N. R. and W. J. Pound. "R and D Project Selection: Where We Stand," *IEEE Transactions on Engineering Management*, 11:124–134 (December, 1964).

8. Cetron, M. J., J. Martino, and L. Roepcke. "The Selection of R&D Program Content—Survey of Quantitative Methods," *IEEE Transactions on Engineering Management*, EM-14:4–13 (March, 1967).

9. Kasner, E. *Essentials of Engineering Economics*. New York:McGraw-Hill Book Company (1979).

10. Disman, S. "Selecting R&D Projects for Profit," *Chemical Engineering*, pp. 87–90 (December, 1962).

11. Kasner, op. cit.

12. Murdick, R. G. and E. W. Karger. "The Shoestring Approach to Rating New Products," *Machine Design*, pp. 86–89 (Jan., 1973).

13. Seiler, R. E. *Improving the Effectiveness of Research and Development*. New York:McGraw-Hill Book Company (1975).

14. Shannon, R. E. *Engineering Management*. New York:John Wiley and Sons (1980).

15. Dean, B. V. and S. S. Sengupta. "Research Budgeting and Project Selection," *IEEE Transactions on Engineering Management*, pp. 158–169 (December, 1962).

16. Mottley, C. M. and R. D. Newton. "The Selection of Projects for Industrial Research," *Operations Research*, pp. 740–751 (1959).

17. Sobelman, S. A. *Modern Dynamic Approach to Product Development*, Picatinny Arsenal, Dover, NJ (December, 1958).

CHAPTER 7

Introduction to Project Management

7.1 INTRODUCTION

This is an important discipline which has emerged essentially since the last World War. The development of complex weapons systems by the United States Defense Department played a significant role in the emergence of this discipline. The project planning, scheduling and control techniques such as Critical Path Method (CPM) and Program Evaluation and Review Technique (PERT), ever since their inception, have also played a vital role in stimulating interest in the discipline of project management.

The history of the Critical Path Method goes back to 1956, when the Integrated Engineering Control Group of E. I. du Pont de Nemours & Company used a network model to schedule activities related to design and construction. This model denoted activities on nodes and the time flow for the precedence relation by the direction of arrows [1]. At that time Du Pont used UNIVAC computers and asked the Remington Rand Corporation to implement the method on the computer. Subsequently, J. W. Manchly and J. E. Kelly jointly refined the method. For the trial run, the CPM was used to construct a $10 million chemical plant in Louisville, Kentucky, in 1957. The CPM network of this plant was managed by six engineers and contained more than 800 activities.

The Program Evaluation and Review Technique was developed at about the same time as CPM. This technique was developed to monitor the effort of 250 prime contractors and 9,000 sub-contractors associated with the U.S. Polaris project. This approach was the result of efforts of a team formed by the U.S. Navy's Special Project Office on January 27, 1958. This team included the members from a consulting company known as Booz, Allen, and Hamilton and from the Lockheed Missile System Division [1]. The technique they developed is believed to be an extended version of the Gantt chart and is very similar to the one developed by Du Pont.

This chapter briefly describes the various aspects of project management with emphasis on Critical Path Method and Program Evaluation and Review Technique.

7.2   NEED FOR PROJECT MANAGEMENT

Before we discuss the need for project management, let us examine the meaning of project management. According to Reference [2], it is essentially a form of organization and a philosophy of action that pinpoints a project manager to fulfill the objectives of a project. Broadly speaking, the following three things are entailed in project management:

   (i) The project has certain objectives
  (ii) The project deadlines are specified
 (iii) The required resources are negotiated by the project manager with top management.

According to Karger and Murdick [2], sometimes project management is also known by the following three names:

   (i) Product management
  (ii) Systems management
 (iii) Program management

The following factors will indicate the need for project management:

   (i) The present organization work schedules will be disrupted with the integration of the project's activities.
  (ii) The project has tight time schedules to meet.
 (iii) The project has to be controlled closely due to penalty contracts.
 (iv) There is a requirement for major innovation.

7.3   CHARACTERISTICS OF A PROJECT MANAGEMENT PROCEDURE

Some of the characteristics of such a procedure are as follows [2]:

    (i) The organization is project oriented.
   (ii) The project has a limited life.
  (iii) Frequent progress evaluation
  (iv) Clearly outlined goal responsibility
   (v) Clearly outlined task and subtask responsibilities
  (vi) Dynamic changes in responsibilities
 (vii) Dynamic changes in organization
(viii) Poor project management control over participants from other companies

7.4   RESPONSIBILITIES OF A PROJECT ORGANIZATION

For a project organization to have total control, the following responsibilities are necessary [3]:

   (i)   Changes associated with the project
  (ii)   Outside contact for the project
 (iii)   Scheduling the project
 (iv)   Monitoring the project
  (v)   Establishing the product definition with respect to hardware, software, and so on
 (vi)   Allocating funds and assigning tasks to groups associated with the project
(vii)   Making decisions on whether the hardware and services for the project are going to be provided by the company or purchased from the outside bodies
(viii)   Investigating the market potentials concerning the project
 (ix)   Identifying those problems which are pertinent to the success of the project and taking necessary measures to overcome such problems
  (x)   Controlling principal subcontractors involved in major tasks

7.5   LIFE CYCLE PHASES OF THE PROJECT ORGANIZATION
AND FUNCTIONS OF PROJECT MANAGEMENT

Both these topics are discussed separately below.

7.5.1   Life Cycle Phases of the Project Organization

The life cycle of a project organization may be divided into six phases [2]. All of these phases are shown in Figure 7.1.

7.5.2   Functions of Project Management

Broadly speaking, the functions of project management are as follows:

   (i)   Project reporting
  (ii)   Planning
 (iii)   Directing
 (iv)   Project execution

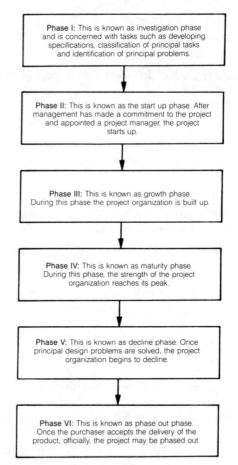

Figure 7.1.
Life cycle phases of the project organization.

  (v) Liaison with customer

 (vi) Evaluating the project

(vii) Reprogramming

All of these functions are self-explanatory. For a description of each, see Reference [4].

7.6 ACTIONS TO STIMULATE PROJECT SUCCESS

This section presents those actions which help to stimulate project success [5]. Some of these actions, which can be taken by the project manager and his team, are stated below:

   (i) Do not over-relay on a specific control tool.

  (ii) Establish realistic objectives.

 (iii) Establish realistic estimates for performance, cost, and schedule.

 (iv) Make sure that a good relationship exists between team and customer.

  (v) Make sure that changes are kept under control.

 (vi) Choose as important members of the team only those persons who have proven track records in their areas of specialization.

 (vii) Seek necessary authority.

(viii) Emphasize that the stated objectives are fulfilled.

 (ix) Emphasize that the stated schedule, performance, and cost estimates are fully satisfied.

  (x) Try to find ways and means by which the effective team members are assured of their job security.

 (xi) Make sure that the structure of the team is flat and flexible.

 (xii) To overcome potential difficulties, make sure of the existence of backup strategies.

(xiii) Try to improve the project's public image.

(xiv) Make sure that a person has the right to choose his own key team members.

7.7 THE PROJECT MANAGER

This section briefly explains the two important aspects concerned with the project manager. These two aspects are the basic responsibilities of the manager and the qualifications of a successful project manager [6]. Thus, both these topics are briefly described below.

7.7.1   Basic Responsibilities of a Project Manager

There are basically three responsibilities which a project manager is expected to fulfill. These are as follows:

(i) To make sure that the end product is delivered to the customer within the framework of a specified time schedule

(ii) To make sure that the end product is produced within the allocated budget

(iii) To make sure that the end product fully satisfies all the performance requirements

7.7.2   Qualifications of Successful Project Manager

A successful project manager usually possesses the following basic qualifications [6]:

(i) Good knowledge of problems of management, for example, law, contracts, marketing, personnel administration, control and purchasing

(ii) Understanding of the concepts of profitability

(iii) A working knowledge of several areas of science

(iv) A strong desire in training, teaching and developing his supervisory manpower

(v) Desirable past experiences

7.8   CRITICAL PATH SCHEDULING TECHNIQUES

This section presents two graphical techniques which are commonly used in planning and controlling projects. These graphical techniques are known as the Critical Path Method (CPM) and Program Evaluation and Review Technique (PERT). The historical aspect of both these techniques is briefly discussed in the introduction to the chapter.

The three pertinent factors of concern in a project are the availability of resource, time and cost. Therefore, both the CPM and PERT approaches can deal with such factors individually and in combination.

Basically, the CPM and PERT approaches are the same. However, their major differences are discussed below [7].

PERT

This is used when there is existence of uncertainties associated with completion times of activities of the project. Therefore, its application is emphasized in research and development work.

CPM

This is used where one is certain about the duration times of activities. Therefore, it is commonly used to manage construction projects.

Both the CPM and the PERT find application in areas such as highway construction, power generation, plant and oil refinery maintenance, shipbuilding and aerospace projects.

A project has to possess specific characteristics for these techniques to be most applicable. These characteristics are as follows [8]:

(i) When a job is started, it has to continue without any interference until its full completion.

(ii) Jobs or tasks are defined such that their completion will result in the end of the project.

(iii) Jobs or tasks are ordered in a specified sequence in which they follow each other.

(iv) Jobs or tasks are independent. In other words, within a defined sequence they can be performed, started and stopped independently.

Both these methods roughly involve the following steps:

(i) Decomposing a project into individual jobs or tasks

(ii) Arranging the individual jobs or tasks into a logical network

(iii) Estimating the time duration of each individual job or task

(iv) Developing a schedule

(v) Finding out those critical jobs or tasks which control the completion of the project

(vi) Redistributing resources or funds so that the schedule is improved

The steps associated with both these techniques are discussed separately below.

7.8.1 Steps to Develop and Analyse a PERT Network

This section presents steps which are usually followed to develop and solve a PERT network [8]. These steps are shown in Figure 7.2. Steps 1 and 2 are self-explanatory. However, the remaining steps are described below.

STEP 3

The PERT network requires three estimates for the duration time of each activity. These estimates may be the judgements of three individual persons. These three estimates are described below:

(i) *The optimistic time:* This is the minimum time an activity or a task will take for its completion.

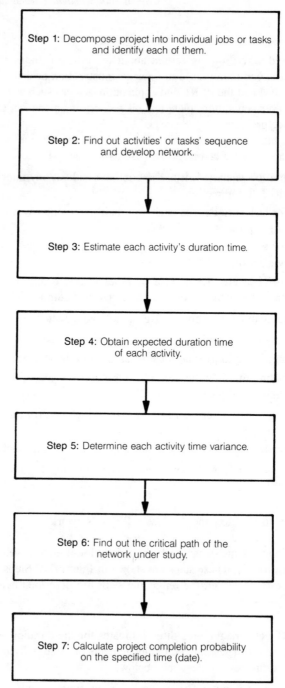

Figure 7.2.
Steps involved in developing and solving a PERT network.

(ii) *The pessimistic time:* This is the maximum time an activity or a task will take for its completion.

(iii) *The most likely time:* This is the most likely time an activity will take for its completion. This is the time which is used for the activities of the CPM network.

STEP 4

An activity or a job expected time duration (i.e., expected time to complete an activity or a task) is given by

$$T_e = \frac{T_0 + 4\,T_m + T_p}{6} \tag{7.1}$$

where

T_e is the expected duration time of an activity or a task.
T_0 is the optimistic duration time of an activity or a task.
T_p is the pessimistic duration time of an activity or a task.
T_m is the most likely duration time of an activity or a task.

Equation (7.1) is based on Beta distribution.

EXAMPLE 7.1

An activity duration time estimates are as follows:

$$T_0 = 20 \text{ days}$$

$$T_p = 30 \text{ days}$$

$$T_m = 24 \text{ days}$$

Calculating the expected duration time of the activity by substituting the given data into Equation (7.1) results in

$$T_e = \frac{T_0 + 4\,T_m + T_p}{6} = \frac{20 + 4(24) + 30}{6}$$

$$= 24.33 \text{ days}$$

STEP 5

This step is concerned with calculating the variance of each activity duration time. The following formula is used to determine the variance:

$$\sigma^2 = \left\{ \frac{T_p - T_0}{6} \right\}^2 \tag{7.2}$$

where σ^2 is the variance of the activity duration time.

EXAMPLE 7.2

An activity optimistic and pessimistic duration time estimates are 50 days and 80 days, respectively. In addition, the most likely duration time estimate for the activity is 60 days. Calculating the variance of the activity time by substituting the given data into Equation (7.2) leads to

$$\sigma^2 = \left\{ \frac{80 - 50}{6} \right\}^2$$

$$= 25$$

STEP 6

The critical path of the network is given by the longest path of that network. The duration time of the project is given by the total sum of the activity expected duration times of the longest path. Other approaches to determine the critical path of the network are discussed later in the chapter. The term "critical" is used to signify that if any delay in the completion of activities along the longest path occurs then the completion of the whole project will be delayed.

STEP 7

This is concerned with calculating the project completion probability on the specified time (date). The following x transformation formula is used:

$$x = (T_d - t_{ec})/\sqrt{s} \tag{7.3}$$

where

s is the total sum of variances of the activity duration times along critical path (i.e., $s = \Sigma \sigma^2$).
t_{ec} is the last network activity's earliest expected completion time.
T_d is the due date for the completion of the project.

Equation (7.3) is associated with the normal probability distribution function. Thus the probability values for the corresponding values of x can be ob-

Table 7.1 Cumulative Normal Distribution Function Tabulated Values.

x	Probability
−4	0.00003
−3.5	0.00023
−3	0.00135
−2.5	0.006
−2	0.02
−1.5	0.07
−1	0.16
−0.5	0.31
0	0.5
0.5	0.69
1.0	0.84
1.5	0.93
2	0.98
2.5	0.99
3	0.999
3.5	0.9998
4	0.9999

tained from Table 7.1. However, for any other values of x, which are not given in this table, the corresponding probability values may be found in a standard probability text book.

EXAMPLE 7.3

The PERT network of a project contains nine activities. After the analysis it is found that the last network activity's earliest expected completion time is 44 days. In addition, the total sum of the variances of the activity duration times along critical path is 16 (i.e., $\Sigma\sigma^2 = 16$). The due date for the completion of the project is 52 days. Calculate the probability that the project will be accomplished on due date or specified time.

In this example, the data are specified for the following items:

$$T_d = 52 \text{ days}, \ s = 16, \ t_{ec} = 44 \text{ days}$$

Thus substituting the above data into Equation (7.3) results in

$$x = (52 - 44)/\sqrt{16}$$

$$= 2$$

From Table 7.1 for $x = 2$, the corresponding probability value is 0.98. Thus the probability of completing the project on the specified date is approximately 98 percent.

7.8.2   Steps to Develop and Analyse a CPM Network

The following steps are associated with the construction and analysis of a CPM network:

    (i) Decompose the project into individual jobs or tasks and identify each of them.

   (ii) Find out activities' or tasks' sequence and develop the network.

  (iii) Estimate each activity's duration time.

  (iv) Find out the critical path of the network under study.

7.9   SYMBOLS AND DEFINITIONS USED TO CONSTRUCT AND SOLVE A CPM OR A PERT NETWORK

All the symbols used to construct either a CPM or a PERT network are shown in Figure 7.3. These symbols are described below [9].

CIRCLE

It is used to represent an event. Broadly speaking the circle represents an unambiguous point in time in the project's life. For example, an event can be a start or a completion of an activity or activities. Usually the events of a CPM or a PERT network are labeled with a number.

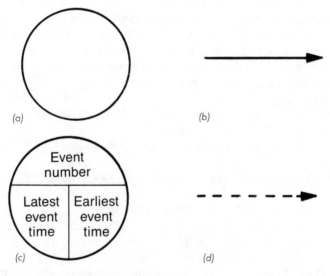

Figure 7.3.
Symbols used to construct a CPM or a PERT network: (a) circle; (b) continuous arrow; (c) circle with divisions; (d) dotted arrow.

CONTINUOUS ARROW

This is <u>used to represent an activity</u>. An activity always begins from a circle and ends at a circle. In addition, to accomplish an activity requires time, manpower and money.

DOTTED ARROW

This represents a dummy activity or a restraint. It is an imaginary activity and does not consume any money, manpower or time. More clearly, it is accomplished in zero time.

CIRCLE WITH DIVISIONS

A circle with divisions also represents an event. However, this circle is divided into three portions as shown in Figure 7.3(c). The top portion is used to label the event with a number. The bottom portion of the circle is divided into two halves. The left half is used for the latest event time whereas the remaining portion is used for the earliest event time. Both the earliest and latest event times are defined below.

Earliest Event Time

This is the earliest time in which an event can be reached; in other words, the earliest time in which an activity can be completed.

Latest Event Time

This is the latest time in which an event can be reached without delaying the completion of the project.

CRITICAL PATH OF THE NETWORK

This is that path which is the longest path from one end (i.e., from the first event) of a CPM or a PERT network to another end (i.e., to last event). For the completion of the project on time, the activities along the critical path have to be accomplished on time; otherwise, the project will be delayed. The total time of the critical path is given by the largest sum of expected activity times of all the paths which originate from the first event and terminate at the last event.

EXAMPLE 7.4

Explain the dummy activity with the aid of a CPM network diagram.

Thus we assume that a small project is described by the CPM network shown in Figure 7.4. The capital alphabetic letters denote activities. Each number in a circle indicates the event number.

In Figure 7.4 the activity C is the dummy activity. It indicates that the activity D can be started only after the completion of both activities A and F. However, the activity B can be started after the completion of activity A.

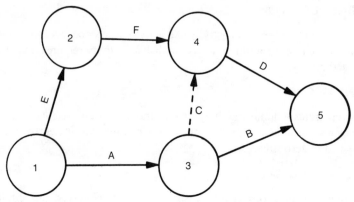

Figure 7.4.
A CPM network.

EXAMPLE 7.5

A small construction project was broken down into eight activities as shown in Table 7.2. Each activity expected duration time is also listed in the table. Construct the CPM network for the project and determine its critical path.

Figure 7.5 shows a CPM network for the data given in Table 7.2. This network is constructed with the aid of symbols given in section 7.9. In this diagram the expected duration time for each activity is shown in brackets.

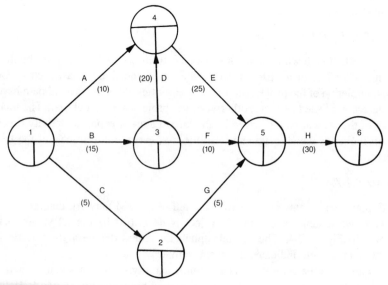

Figure 7.5.
A CPM network for the data specified in Table 7.2.

Table 7.2. Data for Activities of a Project.

Activity	Activity designation	Immediate predecessor activity or activities	Expected duration time for each activity (days)
A	1–4	–	10
B	1–3	–	15
C	1–2	–	5
D	3–4	1–3	20
E	4–5	3–4,1–4	25
F	3–5	1–3	10
G	2–5	1–2	5
H	5–6	4–5,3–5,2–5	30

The network contains four paths which originate from event 1 and terminate at event 6. These paths are as follows:

(i) C–G–H

(ii) B–F–H

(iii) B–D–E–H

(iv) A–E–H

Expected completion times of all the above paths are calculated as follows:

PATH C—G—H

$$\text{Expected completion time of the path} = T_C + T_G + T_H \qquad (7.4)$$

where

T_C is the expected time to complete activity C.

T_G is the expected time to complete activity G.

T_H is the expected time to complete activity H.

Thus substituting the given time data for activities C, G and H into Equation (7.4) results in:

$$\text{Expected completion time of the path (C–H)}$$

$$= 5 + 5 + 30$$
$$= 40 \text{ days}$$

PATH B–F–H

$$\text{Expected completion time of the path} = T_B + T_F + T_H \qquad (7.5)$$

where

T_B is the expected time to complete activity B.
T_F is the expected time to complete activity F.

   Substituting the given time data for activities B, F and H into Equation (7.5) results in:

$$\text{Expected completion time of the path } (B-F-H)$$

$$= 15 + 10 + 30$$

$$= 55 \text{ days}$$

PATH B–D–E–H

$$\text{Expected completion time of the path} = T_B + T_D + T_E + T_H \qquad (7.6)$$

where

T_E is the expected time to complete activity E.
T_D is the expected time to complete activity D.

   Substituting the specified time data for activities B, D, E and H into Equation (7.6) results in:

$$\text{Expected completion time of the path } (B-D-E-H)$$

$$= 15 + 20 + 25 + 30$$

$$= 90 \text{ days}$$

PATH A–E–H

$$\text{Expected completion time of the path} = T_A + T_E + T_H \qquad (7.7)$$

where T_A is the expected time to complete activity A.

Substituting the specified time data for activities A, E and H into Equation (7.7) results in:

Expected completion time of the path (A−E−H)

$$= 10 + 25 + 30$$

$$= 65 \text{ days}$$

Thus the largest expected completion time of the path is given by path B−D−E−H. This time is equal to 90 days. Therefore this path is the critical path.

7.10   ESSENTIAL FORMULAS AND A PROCEDURE FOR DETERMINING THE CRITICAL PATH OF A NETWORK

This section presents two important topics. Both these are discussed separately below.

7.10.1   Essential Formulas

This section presents four formulas essential to performing network analysis. These formulas are defined with the aid of Figure 7.6. The symbology used in Figure 7.6 is defined below.

$t_1(j)$ is the latest event time of the event j.
$t_2(j)$ is the earliest event time of the event j.
$t_1(k)$ is the latest event time of the event k.
$t_2(k)$ is the earliest event time of the event k.
$y(j,k)$ is the expected completion time of the activity between events j and k.

Thus for activity (j,k) we have

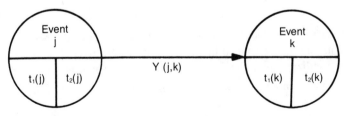

Figure 7.6.
A single activity network.

FORMULA A

$$\text{Latest start time} = t_1(k) - y(j,k) \qquad (7.8)$$

where

$t_1(k)$ is the latest event time of the event k.
$y(j,k)$ is the expected completion time of the activity between events j and k.

FORMULA B

$$\text{Latest finish time} = t_1(k) \qquad (7.9)$$

FORMULA C

$$\text{Earliest finish time} = t_2(j) + y(j,k) \qquad (7.10)$$

where $t_2(j)$ is the earliest event time of event j.

FORMULA D

$$\text{Earliest start time} = t_2(j) \qquad (7.11)$$

FORMULA E

$$\text{Total float} = t_1(k) - t_2(j) - y(j,k) \qquad (7.12)$$

7.10.2   A Procedure for Determining the Critical Path of a Network

This is basically a four-step approach. These steps are described below.

(i) Construct network.

(ii) Determine the earliest event time, t_2, of each event of the network. This can be determined by making a forward pass of the network and utilizing the following formula, where k represents any event:

$t_2(k)$ = maximum for all proceeding event j of

$$[t_2(j) + y(j,k)] \qquad (7.13)$$

$$t_2 \ (\text{first event}) = 0 \qquad (7.14)$$

(iii) Determine the latest event time, t_1, of each event of the network. This can be determined by making a backward pass of the network and utilizing the following formula, where j represents any event:

$t_1(j)$ = minimum for all succeeding event k of

$$[t_1(k) - y(j,k)] \qquad (7.15)$$

$$t_1(\text{last event}) = t_2(\text{last event}) \qquad (7.16)$$

(iv) Choose those network events whose $t_1 = t_2$. If the network has only one path with event satisfying the condition $t_1 = t_2$, then such a path is critical. However, if there is more than one such path satisfying the $t_1 = t_2$ condition, then go to next step.

(v) For each path which satisfies the condition of step (iv), determine the total floats for all its activities. Sum the total floats of each path's activities. The path which has the least sum of the total floats is the critical path of the network.

EXAMPLE 7.6

For the network shown in Figure 7.5, determine the critical path by calculating each event's earliest and latest event times.

To perform earliest and latest event times analysis, the Figure 7.5 network is redrawn in Figure 7.7. With the aid of Equations (7.13)–(7.14) and by making a forward pass, we calculated the earliest event times of network shown in Figure 7.7.

For example, the earliest event time of event 4 was determined by making the forward pass. Thus event 4 has two incoming passes, i.e., through activities 1–4 and 3–4. Therefore from activity 1–4 we obtain:

$$t_2(4) = t_2(1) + y(1,4) = 0 + 10 = 10 \text{ days}$$

and from activity 3–4:

$$t_2(4) = t_2(3) + y(3,4) = 15 + 20 = 35 \text{ days}$$

Thus with the aid of Equation (7.13), we select 35 days as the earliest time for event 4. Similarly, with the aid of Equations (7.12) and (7.16) and by making a backward pass, we calculated the latest event times of network shown in Figure 7.7.

For example, the latest event time of event 3 was determined by making the backward pass. Thus event 3 has two incoming backward passes, i.e., through activities 3–4 and 3–5.

Therefore, from activity 3–4 we obtain:

$$t_1(3) = t_1(4) - y(3,4) = 35 - 20 = 15 \text{ days}$$

and from activity 3–5:

$$t_1(3) = t_1(5) - y(3,5) = 60 - 10 = 50 \text{ days}$$

Thus with the aid of Equation (7.17), we select 15 days as the latest event time for event 3.

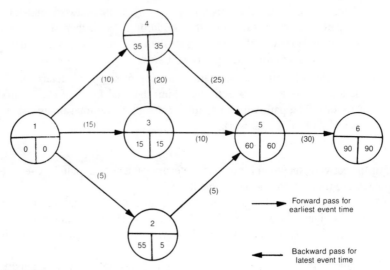

Figure 7.7.
A CPM network with earliest and latest event times.

In Figure 7.7, there are three paths each of whose event's earliest and latest event times are equal. These paths are as follows:

(i)   Path 1−4−5−6
(ii) Path 1−3−4−5−6
(iii) Path 1−3−5−6

Therefore, to determine the critical path, we have to calculate the total float of each activity of these three paths. With the aid of Equation (7.12), the total floats of all the three paths' activities are calculated as follows:

ACTIVITY 5–6:

$$\begin{aligned}
\text{Total float} &= t_1(6) - t_2(5) - y(5,6) \\
&= 90 - 60 - 30 \\
&= 0 \text{ days}
\end{aligned}$$

ACTIVITY 3–5:

$$\begin{aligned}
\text{Total float} &= t_1(5) - t_2(3) - y(3,5) \\
&= 60 - 15 - 10 \\
&= 35 \text{ days}
\end{aligned}$$

ACTIVITY 4–5:

$$\begin{aligned}
\text{Total float} &= t_1(5) - t_2(4) - y(4,5) \\
&= 60 - 35 - 24 \\
&= 0 \text{ days}
\end{aligned}$$

ACTIVITY 3–4:

$$\begin{aligned} \text{Total float} &= t_1(4) - t_2(3) - y(3,4) \\ &= 35 - 15 - 20 \\ &= 0 \text{ days} \end{aligned}$$

ACTIVITY 1–3:

$$\begin{aligned} \text{Total float} &= t_1(3) - t_2(1) - y(1,3) \\ &= 15 - 0 - 15 \\ &= 0 \text{ days} \end{aligned}$$

ACTIVITY 1–4:

$$\begin{aligned} \text{Total float} &= t_1(4) - t_2(1) - y(1,4) \\ &= 35 - 0 - 10 \\ &= 25 \text{ days} \end{aligned}$$

Thus the sum of total floats for paths (i)–(iii) are as follows:

PATH 1–4–5–6

Sum of total floats = (Float of activity 1–4) + (Float of activity 4–5) + (Float of activity 5–6) = 25 + 0 + 0 = 25 days

PATH 1–3–4–5–6

Sum of total floats = (Float of activity 1–3) + (Float of activity 3–4) + (Float of activity 4–5) + (Float of activity 5–6) = 0 + 0 + 0 + 0 = 0 days

PATH 1–3–5–6

Sum of total floats = (Float of activity 1–3) + (Float of activity 3–5) + (Float of activity 5–6) = 0 + 35 + 0 = 35 days

The least sum of total floats out of the above three paths is given by the path 1–3–4–5–6. Therefore, the path 1–3–4–5–6 is the critical path of the network.

EXAMPLE 7.7

A small research and development project is composed of six activities. These activities along with their estimated times are identified in Table 7.3. Construct the PERT network for the project and determine the project probability of completion on time. The project due date is 54 days.

For the specified data, the Figure 7.8 network was constructed. Earliest and latest event times for each event were calculated with the aid of Equations (7.13) and (7.15), respectively. These times are shown in Figure 7.8.

The critical path of the network shown in Figure 7.8 is given by the longest sequence of connected activities through the network, i.e., A–C–E–F.

Table 7.3. Time Estimates for Activities.

Activity	Activity designation	Immediate predecessor activity or activities	Time estimates for each activity (days)			Expected times $t_e = \dfrac{t_0 + 4T_m + T_p}{6}$	Variance, σ^2, for each activity $\sigma^2 = \left\{\dfrac{T_p - T_0}{6}\right\}^2$
			t_0	T_m	t_p		
A	1-3	—	20	25	30	25	2.78
B	1-2	—	10	15	20	15	2.78
C	3-2	1-3	8	10	12	10	0.444
D	3-4	1-3	10	10	10	10	0
E	2-4	1-2, 3-2	10	12	14	12	0.444
F	4-5	3-4, 2-4	5	5	5	5	0

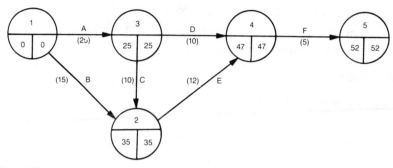

Figure 7.8.
The PERT network.

Thus the total of the variances along critical path from Table 7.3 is

$$S = 2.78 + 0.444 + 0.444 + 0$$
$$= 3.668$$

Substituting the specified data into Equation (7.3) results in:

$$x = \frac{T_d - t_{ec}}{\sqrt{s}} = \frac{54 - 52}{\sqrt{3.668}} = 1.044 \cong 1.0$$

For the above calculated value, from Table 7.1 the probability of completing the project on time is about 84 percent.

7.11 BENEFITS AND DRAWBACKS OF THE CRITICAL PATH METHOD

From Reference [10], some of the benefits and the drawbacks of the critical path method are as follows.

BENEFITS

(i) It helps to improve communication and understanding.
(ii) It helps in cost control and cost savings.
(iii) It depicts interrelationship in work flow.
(iv) It identifies those work activities which are vital to completing the project on time.
(v) Needs for labour and resources can be determined in advance.
(vi) It can determine the duration of the project systematically.
(vii) It can be computerized.

(viii) It helps to avoid duplications and omissions.

(ix) It helps to monitor the progress of the project effectively.

(x) It allows alternative simulations.

DRAWBACKS

(i) Costly and time-consuming

(ii) Poor time estimates

(iii) Bias to use pessimistic time estimates

7.12 SUMMARY

This chapter briefly discusses the subject of project management with emphasis on project planning, scheduling and control techniques such as Critical Path Method (CPM) and Program Evaluation and Review Technique (PERT).

The chapter begins by briefly outlining the historical aspects of project management. The need for project management is briefly discussed along with the characteristics of project management procedure. In addition, the responsibilities of the project organization are listed and its life cycle phases are briefly discussed. The life cycle project organization is categorized into six phases. In addition, the chapter goes on to list the actions necessary to stimulate project success. The basic responsibilities and qualifications of a project manager are discussed.

The emphasis of the chapter is on critical path scheduling techniques. Thus both the Program Evaluation and Review Technique (PERT) and Critical Path Method (CPM) are described in detail. The steps associated with both these techniques are outlined. Seven steps are involved in developing a PERT network. Similarly, four steps are involved in developing a CPM network. In addition, essential formulas concerning both CPM and PERT and a procedure to determine the Critical Path of a network are presented. The advantages and disadvantages of the Critical Path Method are listed. The chapter contains seven examples along with their solutions.

7.13 EXERCISES

1. What are the factors which may indicate the need for project management?

2. What are the responsibilities of a project organization?

3. Describe the life cycle phases of a project organization.

4. Discuss the historical aspects of both the Critical Path Method and Program Evaluation and Review Technique.

5. What are the responsibilities and attributes of a project manager?

6. Discuss those actions which may be taken to stimulate project success.

7. What are the differences between the CPM and PERT?

8. What are the steps to be followed to develop a PERT network?

9. A PERT network activity's optimistic, most likely and pessimistic duration time estimates are 42, 56 and 60 days, respectively. Calculate the expected duration time of the activity.

10. What is the dummy activity?

11. A CPM network of a project is shown in Figure 7.9. Duration time estimate (days) for each activity is specified in the diagram. Determine
   (i) the earliest and the latest event times for each event.
   (ii) the critical path of the network.

12. A small research and development project is broken down into seven activities as shown in Table 7.4. The optimistic, most likely and pessimistic duration times, respectively, for each activity are also specified in Table 7.4. The project due date is 40 days. Construct the PERT network for the project, and, in addition, the project probability of completion on time.

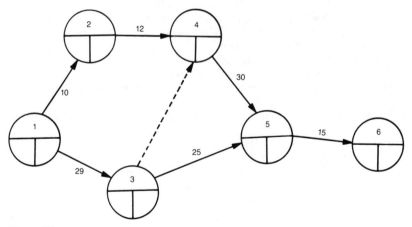

Figure 7.9.
A CPM network.

Table 7.4. Time Estimates for Project Activities.

Activity	Activity designation	Immediate predecessor activity or activities	Time estimates for each activity (days)		
			t_0	t_m	t_p
A	1–2	—	10	10	10
B	1–3	—	7	10	15
C	2–4	1–2	8	12	16
D	3–5	1–3	15	20	25
E	4–5	2–4, 3–4	7	7	7
F	5–6	4–5, 3–5	4	5	6
G	3–4	1–3	9	11	20

7.14   REFERENCES

1. Riggs, J. L. and M. S. Inoue. *Introduction to Operations Research and Management Science: A General Systems Approach.* New York:McGraw-Hill Book Company (1975).
2. Karger, D. W. and R. G. Murdick. *Managing Engineering and Research.* 200 Madison Ave., New York 10016:Industrial Press Inc. (1969).
3. Middleton, C. J. "How to Set Up a Project Organization," *Harvard Business Review,* 73–82 (March/April 1967).
4. Martin, C. C. *Project Management: How to Make it Work.* New York:AMA COM: A Division of American Management Associations (1976).
5. Kerzner, H. *Project Management: A Systems Approach to Planning, Scheduling and Controlling.* New York:Van Nostrand Reinhold Company (1979).
6. Gaddis, P. O. "The Project Manager," *Harvard Business Review,* 89–97 (May/June 1959).
7. Lee, S. M., L. J. Moore and B. W. Taylor. *Management Science,* Dubuque, IA:Wcb: Wm. C. Brown Company, Publishers (1981).
8. Chase, R. B. and N. J. Aquilano. *Production and Operations Management: A Life Cycle Approach.* Homewood, IL 60430:Richard D. Irwin, Inc. (1981).
9. Dhillon, B. S. *Reliability Engineering in Systems Design and Operation.* New York:Van Nostrand Reinhold Company (1983).
10. Lomax, P. A. *Network Analysis: Applications to the Building Industry.* London:The English Universities Press Limited (1969).

CHAPTER 8

Management of Technical Proposals and Specifications

8.1 INTRODUCTION

In engineering business the understanding and management of technical proposals and specifications is very important. Every engineer or engineering manager gets involved directly or indirectly with both these items.

The proposal is used by the engineering organization to bring in new business which will ultimately make profit for the company. Therefore, such a document has to be prepared in such a way that it is able to win the contract for the company at the most profitable terms and conditions. If the personnel of a company is unable to produce an effective proposal then there are high chances that the firm may not be able to remain in business for a long time. According to References [1] and [2], the proposal is described as follows. It is a document comprising an offer from the contractor (or bidder) to the customer (or owner) to carry out a certain task by a specified time at the defined level of quality for a certain amount of money with the aid of his manpower and facilities.

Broadly speaking, the specifications are used to convey clearly and accurately the information regarding the work to be performed. Therefore the importance of specifications is very clear. For example, an inaccurate specification will lead to the wrong results of the work or the task. The specifications also serve the following purposes [2]:

(i) They compare bidders for the work in question.

(ii) They ensure the customer gets appropriate return on the investment.

(iii) They provide a tool to the customer to ensure legality, safety and so on in the accomplished job.

This chapter describes the various important aspects of technical proposals and specifications separately.

8.2 TECHNICAL PROPOSALS

As mentioned earlier, the proposal is an offer, made by the contractor to the prospective customer, to perform specific work. In engineering business the proposals are used regularly to bid for contracts. Therefore, an organization must organize itself properly to prepare effective proposals in order to secure contracts in the competitive market. For that reason, this section presents the important aspects of technical proposals, as follows.

8.2.1 Types of Technical Proposals

According to Reference [3], the proposals may be classified into the following categories.

SOLICITED TECHNICAL PROPOSAL

This type of proposal is submitted by the contractor to customer for the work to be done after receiving a request for proposal from the customer. The request for proposal contains the information on work to be accomplished. Furthermore, it may also contain information on presentation of the needed information from the prospective contractor. Usually the customer requires information from the contractor such as follows:

(i)   Plans of the contractor to attack the work in question.

(ii) Contractor's qualifications.

(iii) Price of the contractor's effort.

(iv) Work schedule.

UNSOLICITED TECHNICAL PROPOSAL

The proposal which is initiated by the contractor without a formal request for proposal from the customer is known as the unsolicited proposal. This type of proposal is used in a situation, for example, when the contractor's organization feels that there is a definite need for specific work (job) or a study and such work or study needs financing from an outside body. Furthermore, the contractor may learn informally that there is a need for submitting a proposal for certain work or study. Sometimes through this route the research funding is secured from sources such as a government organization.

TECHNICAL BROCHURE

According to Reference [3], the technical brochure may be classified as the third type of technical proposal. Furthermore, it may be called the "general proposal." Usually, the technical brochure is used to explain contractor's capabilities to potential customers so they can make use of services offered by the contractor. Therefore, it may be said that the technical brochure is essentially a promotional tool.

Before preparing the technical brochure, careful consideration must be given to such points as audience, purpose, content, cost, style and format.

8.2.2 Upper Management Considerations in the Development of a Proposal

This section lists some of the considerations to be taken into account by the contractor's upper management before preparing a proposal for new work. They are as follows [4]:

(i) Profit.

(ii) Competition.

(iii) Is the work involved in the proposal compatible with company policy?

(iv) If the firm's bid is unsuccessful, will it effect the company image?

(v) What are the requirements for resources?

(vi) Price of the contract.

(vii) What are the risks involved with the project?

(viii) What will be the preparation cost of the proposal?

(ix) Should the company bid low to win the contract?

(x) What is the importance of this project to the company?

(xi) Is there any legal complication involved with the project?

(xii) What will be the duration of this contract?

Now we should appreciate that answering all the questions listed above requires input from the various groups of the contractor's organization. Examples of such groups are cost accounting, contracting, technical and marketing.

8.2.3 A Procedure for Preparing Effective Engineering Proposals

Usually each company has its own approach to developing proposals for new projects. However, the overall approach employed by these companies is somewhat the same. Therefore this section presents a general or a typical procedure for writing effective engineering proposals. This procedure is composed of eight steps as shown in Figure 8.1. All these steps are described briefly as follows [2].

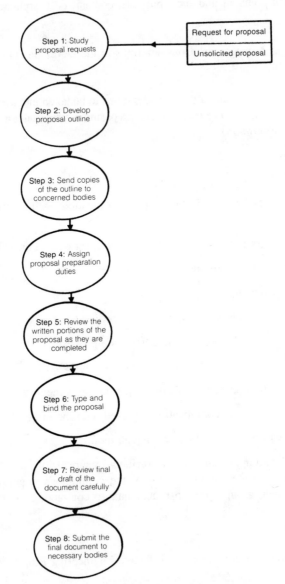

Figure 8.1.
An eight step approach to preparing proposals.

STEP 1

This step involves the investigation of a proposal request or the study of the content of an proposed unsolicited proposal. The objective of this investigation is to thoroughly understand what is really required in the proposed project and to summarize it into a single sentence if possible. This way the objective of the proposal will be clearly understood by the concerned people.

STEP 2

This step is concerned with preparing the proposal outline. It is always wise to prepare the outline of the proposal first, before starting to write the actual proposal. The contents of the outline include topics such as summary, problem under consideration, manpower and facilities. The outline provides a useful tool to guide the thinking of the people involved in the preparation of the proposal.

STEP 3

In this step, the outline of the proposal is circulated to concerned bodies for their suggestions and comments. Once all the comments and suggestions are received, they are reviewed. Finally, the outline is revised to incorporate the useful comments and suggestions.

STEP 4

After the final outline of the proposal is completed, then the next step of the proposal writing procedure is executed. In this step the proposal writing or preparing duties are assigned to groups or persons. Special care must be given when assigning the duties of writing the summary and the problem requiring solution. According to Reference [2], these two items are very important in each proposal. Therefore, the persons must be scrutinized properly when assigning such tasks.

STEP 5

This step is concerned with reviewing the already written portions of the proposal. This should be done as the sections of the proposals are being completed. This way one will have effective control of the work in progress. If there is any departure from the correct path, it will be detected and corrected immediately before too much time is wasted on the incorrect path. When reviewing the written material the attention must be given to points such as clarity, technical accuracy and sequence of the already written material.

STEP 6

This step is concerned with getting the written proposal document typed. Proper care must be given when getting such a document typed, otherwise it may reflect on the company's image.

STEP 7

After the proposal document is typed and is available in the final draft copy, the next step is to review it very carefully to identify any error or weakness. In addition, review the cost figures used in the proposal to see their competitiveness with other bidders, the accuracy and so on.

STEP 8

Submit the necessary copies of the proposal to concerned bodies.

8.2.4   Pertinent Items for Inclusion in Proposals

According to Reference [5], when writing a proposal, one must consider various items for inclusion. Some of them are as follows:

   (i) Purpose, scope and the problem definition.
  (ii) Problem history, need for solution and solution benefits.
 (iii) Past experience, chances of success in obtaining a solution to the problem, and availability of facilities.
  (iv) Techniques or procedures for solving the problem.
   (v) Time required to accomplish the project.
  (vi) Breakdown of project tasks.
 (vii) Price of the project.
(viii) Earlier association reference.
  (ix) Project manpower.
   (x) Qualifications of the involved manpower.
  (xi) References.
 (xii) Proposal specifications.

Furthermore the contractor should give careful consideration to the timing of the proposal, overall appearance of the proposal, over committal in the proposal, format of the proposal, and so on.

8.2.5   Typical Format of a Proposal

This section presents the typical format of a technical proposal [3]. Actual format of the proposal documentation may differ from one project to another, however, essentially the basic format for all the proposals is the same. A typical format of a proposal is given in Figure 8.2.

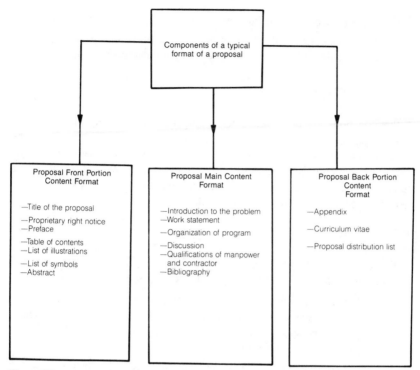

Figure 8.2.
Components of a typical format of a proposal.

8.2.6   Role of Customer Relations in Proposal Preparation

Relationship between customer and contractor plays an important role in the success of engineering proposals. The contractor writes engineering proposals for customers in order to win contracts under most desirable terms and conditions. Therefore it is very important for the contractor to thoroughly understand his customer, especially when submitting the unsolicited proposals. In order to do so, the contractor must have knowledge of the customer's product line, facilities, strengths, weaknesses, important people, operating habits, advance planning, suppliers and so on. According to Reference [6] the following points are helpful in promoting good relations with the customer:

(i) Always remember when writing an engineering proposal that it is being written for people who have a technical background. Therefore, avoid taking the approach of an advertising man.

(ii) Write the proposal well. It must be logical, concise, clear and specific.

(iii) Have enthusiasm when preparing a proposal.

(iv) Have respect for your customer.
(v) Have associated with the proposal only those people who are compatible with their counterparts in the customer's organization.

8.3   ENGINEERING SPECIFICATIONS

Specifications are used when developing new engineering products. The size of specifications depends on the product in question. The quality of the specification will directly affect the product it specifies. Therefore, in the preparation of engineering specifications thoughtful and careful consideration must be given to produce the desirable end product. Various aspects of engineering specifications are described in this section.

8.3.1   Engineering Specifications Classifications

There are various types of specifications. However, for our purpose, we will consider only the following three types of general specifications [2]:

(i) *Specifications type I:* These specifications are concerned with performance and are used to describe the final product requirements. However, the specifications do not show the procedure of how the requirements will be fulfilled.
(ii) *Specifications type II:* These specifications are concerned with design and contain information on how the final product requirements are to be fulfilled.
(iii) *Specifications type III:* These are known as the construction specifications. They describe requirements for items such as materials, legality and techniques needed in actual manufacture of an engineering product.

8.3.2   Specification Writing Hints

This section briefly covers the useful points to be considered when writing a specification [7]. These points are illustrated in Figure 8.3.

8.3.3   Engineering Specifications Layout

This section presents the general format of engineering specifications. However, the specifications layout may vary slightly for different situations. Thus according to Reference [8], the general format of specifications is as follows:

(i) Title sheet: This sheet is used to present information such as the title, name and address of the organization in question, date, etc.
(ii) Table of contents.

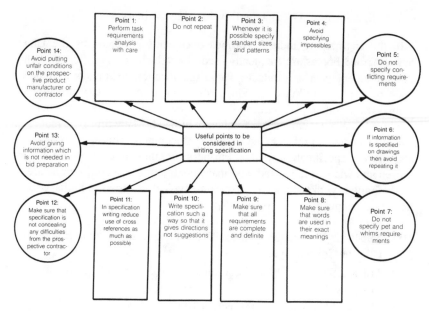

Figure 8.3.
Useful points to be considered when writing specification.

(iii) Introduction to Specification: This section describes the background information associated with the specification under consideration. Information contained in this section is of a general nature.

(iv) Scope: This provides precise information on the product under consideration and its role. Furthermore, it accurately outlines the requirement boundaries.

(v) Other documents associated with specifications.

(vi) Technical Specification: This is the backbone of the overall specification document. It covers the information useful for obtaining effective control of items such as construction, performance and process. The requirements presented in this section are generally divided into various subgroups. Examples of these subgroups are design and construction, quality assurance, materials, performance, reliability, manufacture ratings, dimensions, spares, acceptance conditions, maintainability, maintenance, and so on.

(vii) Particulars of other necessary information: This section covers information on documentary materal. Examples of such documentary material are product maintenance drawings, test reports, and operation handbooks.

(viii) Appendix, Index and Miscellaneous Information: These three items are separately presented in the specification document.

8.3.4   Specification Documents Developed by Military

In the engineering industry various documents produced by the United States Armed Forces are frequently used as a vehicle for specifications when developing engineering products. These specification documents were developed by the U.S. military to reduce confusion in the military procurement from various different contractors. Today, even when civilian products are produced, these documents are frequently utilized. Many times the main specification document refers to these military documents. This helps to save time and effort of specification writers. Furthermore, these documents are already proven and widely recognized, and many contractors or manufacturers are already familiar with such documents. Some typical examples of military specification documents are as follows [9]:

1. Mil-R-22973: Reliability Index-Determination for Avionic Equipment Control General Spec.
2. Mil-R-22732: Reliability Requirements for Shipboard Electrical Equipment.
3. Mil-M-6867: Magnetic Inspection Units.
4. Mil-Q-9858: Quality Program Requirements.
5. Mil-L-10547A: Liners, Case Waterproof.
6. Mil-STD-105D: Sampling Procedures and Tables for Inspection by Attributes.
7. Mil-P-116F: Preservation Packaging, Method of.
8. Mil-Std-1472B: Human Engineering Design Criteria for Military System, Equipment and Facilities.
9. Mil-Std-108E: Definitions of and Basic Requirements for Enclosures for Electric and Electronic Equipment.

These documents are available from the Naval Publications and Forms Center, 5801 Tabor Ave., Philadelphia, PA 19120.

8.3.5   Advantages and Disadvantages of Engineering Specifications

Just like anything else, the engineering specifications have advantages and disadvantages. Some of the advantages are that they (1) help standardization, (ii) help to produce better quality product, and (iii) help to better understand the exact needs of the customer. On the other hand, the disadvantages of specifications are that they are time consuming, may restrict innovation, and so on. Advantages and disadvantages concerning specifications are discussed in Reference [10].

ADVANTAGES OF STANDARD SPECIFICATION FOR MATERIALS

Frequently the standard specifications for materials are used. Some of the advantages of standard specifications are as follows [11]:

(i) Helpful in comparing bids.

(ii) Due to mass production of standardized items, the cost decreases.

(iii) They can be included by reference.

(iv) Reasonably available materials in the market can be chosen by the designer.

(v) A customer can make use of a specification which is already tried.

(vi) Helpful in comparing results of testing carried out by the different organizations.

(vii) Helpful in reducing the manufacturer's stock varieties.

(viii) Provide means of mutual understanding between the customer and the contractor or manufacturer (because the standard specifications usually comprise the thinking of both manufacturers and consumers).

8.4 SUMMARY

This chapter briefly covers the topics of technical proposals and specifications. The chapter begins by briefly introducing both of these topics and then penetrates into the subject of technical proposals. Thus, the three types of technical proposals are described. These are solicited, unsolicited and technical brochure. The next topic concerned with the proposal is the upper management considerations in the development of a proposal. Under this subject an extensive list of considerations are presented.

An eight step procedure for preparing effective engineering proposals is described in detail. The next two items covered under the technical proposal section are items to be considered for inclusion in the proposal and the layout of the proposal, respectively. The role of customer relations in proposal preparation is discussed, and useful hints for promoting good relations with the customer are outlined.

The other main subject of the chapter is engineering specifications. The three types of engineering specifications are briefly described. Furthermore, the hints to be considered in writing specifications are briefly outlined. A general layout of engineering specifications is described. Finally, military specifications and the advantages and disadvantages of engineering specifications are covered briefly.

8.5 EXERCISES

1. Describe what is meant by technical proposal and specifications.
2. What are the principal elements of an engineering proposal?
3. What are the principal purposes of specifications?
4. Describe the following items:
    (i) Technical brochure
    (ii) Material specifications
    (iii) Design specifications
    (iv) Solicited proposal
    (v) Construction specifications
5. What are the main considerations to be given when writing
    (i) An engineering proposal
    (ii) A technical specification
6. Explain the layout of a general proposal.
7. Describe the term "specifications for workmanship."
8. Prepare a specification document for an electric switch.
9. What are the benefits of standard specifications?
10. Write an essay on military specifications.
11. Describe a simplified approach to preparing engineering proposals and specifications.
12. What is meant by "Request for Proposal?"

8.6 REFERENCES

1. Dunham, C. W. *Contracts, Specifications and Law for Engineers*. New York:McGraw-Hill Book Company (1971).
2. Hicks, T. G. *Successful Engineering Management*. New York:McGraw-Hill Book Company (1966).
3. Reisman, S. J., ed. *A Style Manual for Technical Writers and Editors*. New York:The Macmillan Company (1962).
4. Walton, T. F. *Technical Manual Writing and Administrations*. New York:McGraw-Hill Book Company (1968).
5. Houp, K. W. and T. E. Pearsall. *Reporting Technical Information*. New York:Glencoe Press, A Division of Benziger Bruce & Glencoe, Inc. (1973).
6. Clarke, E. "Ten Steps to Better Engineering Proposals," in *Management Guide for Engineers and Technical Administrators*. N. P. Chironis, ed. New York:McGraw-Hill Book Company (1967).
7. Abbett, R. W. *Engineering Contracts and Specifications*. New York:John Wiley & Sons, Inc. (1967).

8. "A Guide to the Preparation of Engineering Specifications," London:The Design Council (1980).
9. Dhillon, B. S. *Reliability Engineering in Systems Design and Operation.* New York:Van Nostrand Reinhold Company (1982).
10. Whittemore, H. L. *Ideas on Specifications.* Connecticut:Columbia Graphs (1952).
11. Hayes, G. E. and H. G. Romig. *Modern Quality Control.* Encino, California:Bruce, A Division of Benziger Bruce & Glencoe, Inc. (1977).

CHAPTER 9

Management of Engineering Contracts

9.1 INTRODUCTION

Each day in the engineering industry many new contracts are signed to develop new products, to sell already manufactured items, and so on. In the engineering industry the contracts are important because they provide information such as the starting date of the project, expectation from the parties which sign the contract, methods of payment for the services, and project completion date. According to Reference [1], the simple definition of a contract is that it is an agreement enforceable by law. The signing parties are bound by the contracted terms. As the laws may vary from state to state, province to province, and country to country, the law governing a certain contract will be determined by the place where the contract was originally signed. However, otherwise it has to be clearly stated in the contract document regarding the governing law. Furthermore, it is emphasized here that not all the contracts are enforceable through the law. In order to be enforceable through the courts, usually a contract must possess specific elements. These elements are as follows [1]:

(i) The contract must be set forth, with regard to form, according to the provisions of the law.

(ii) The contracting parties must be competent.

(iii) A valid consideration must be given for all parties to the contract.

(iv) For the well-being of the contract there must be a real agreement among the contracting parties to the terms specified in the contract.

(v) The contract must possess lawful subject matter.

Note that throughout the chapter the terms contractor and customer are frequently used. The term contractor applies to an organization which manufactures products, constructs buildings, performs services, and so on. On the other hand, the term customer applies to an organization which uses the services of the contractor. These services are concerned with the manufacture of products, construction of buildings, etc.

This chapter briefly presents the various aspects of engineering contracts.

9.2   ESSENTIAL PROVISIONS OF A CONTRACT

This section briefly describes the important provisions in a contract as recommended by the American Society of Civil Engineers. According to this society, as given in Reference [1], any engineering services contract should have provisions for the following essential items:

   (i) Statement on starting and completion time of contract work

   (ii) Date when the contract agreement is signed

   (iii) Contracting parties' names, addresses and descriptions

   (iv) Description of the engineer's obligation to the client

   (v) Summary of the engineer's services scope

   (vi) Payments for the work to be performed

   (vii) Compensation for termination of service prior to the actual completion date

   (viii) Provisions for cancellation of service prior to the actual completion date

   (ix) Plans or copyrights reuse

   (x) Compensation for additional work

9.3   ENGINEERING CONTRACT DOCUMENTS

In a real life situation there are various documents which are associated with a contract, e.g., letters, specifications and so on. Therefore, a properly written contract should state which documents to be regarded as the contract documents. According to Reference [2], the contract documents usually consist of some or all of the items specified in Figure 9.1.

Most of the items given in Figure 9.1 are briefly discussed in the following sections:

   (i) *The Tender:* This forms the "offer" given by the contractor for the customer's eventual acceptance without any condition. The contract conditions to which the offer is subjected are specified in the tender. Furthermore, the tender also specifies items such as the terms and the price of the contract, and the payment method.

   (ii) *Contract Conditions:* These are simply those rules by which the contract is governed. Various standard conditions of contract are available in the market, and these conditions may vary from one contract to another. It is not uncommon to see that some

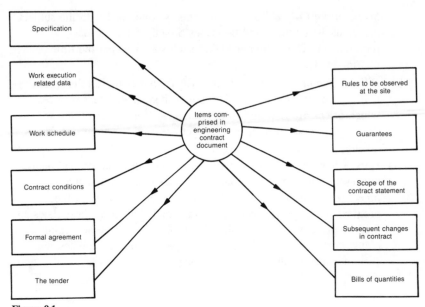

Figure 9.1.
Items comprised in contract documents.

organizations prepare their own conditions of contract which fit best to their situations.

(iii) *Specifications:* This describes the work's description and design information, test procedures, performance standards to be met, maintenance requirements, and so on.

(iv) *Work Execution Related Data:* These are those data which will affect the execution of the work. Examples of such data are constraints on work hours, site conditions, geological data, etc. However, these types of data depend upon the nature of the project under consideration. These types of data are supplied by the customer along with the invitation to tender.

(v) *Rules to be Observed at the Work Site:* These rules are imposed by the customer on the contractor and may concern fire safety, labour relations, cleanliness of site, noise, and so on.

(vi) *Formal Agreement:* This is concerned with the confirmation of a contract in an official agreement and has various points such as summarizing contract highlights, emphasizing the contract importance, etc. The agreement usually displays the principal features of the contract such as finishing dates, payment terms, price of the contract, a list of contract documents, etc.

(vii) *Scope of the Contract Statement:* This is concerned with the subject matter of the contract and describes broadly the customer requirement. The customer makes such statements at the time of inviting tender.

(viii) *Bills of Quantities:* The customer usually issues these bills when inviting tender. One usage of bills is to itemise different services for which prices are required from tenderers. The specifications and drawings associated with the contract are used to obtain customer's quantity estimates.

(ix) *Work Schedule:* This is concerned with recording the critical dates during the lifetime of the contract in question. Some of the objectives of these dates are that the customer is aware of the progress of the project and the contractor is bound by these dates so that the project is accomplished in an orderly fashion according to plans. Furthermore, if there is any slippage from these dates, the customer may be able to claim damage.

9.4   CLASSIFICATIONS OF CONTRACTS

This section briefly presents the classifications of contracts and the factors used to determine such classifications. These are as follows.

9.4.1   Contracts Classification Determining Factors

In the real life environments, there are various types of contracts which can be used to procure engineering products. However, the usage of a particular type of contract for a specific procurement is dictated by various influencing factors such as follows [3]:

(i) Competition for the contract: For example, if there are more firms interested in the project to be contracted then the customer may be able to dictate a specific type of contract which is more beneficial to him.

(ii) Risk: This is another determining factor for the type of contract to be signed. For example, if a project under consideration requires a significant amount of research to accomplish it successfully then the contractor has to assume a considerable amount of risk. This type of situation calls for a specific type of contract to reduce the contractor's risk.

(iii) The type of equipment to be procured.

(iv) Design complexity of the equipment.

(v) Project urgency.

(vi) Contractor's past performance.

(vii) Earlier dealing with the same contractor.

(viii) Difficulty in estimating procurement cost.

(ix) Accounting procedures used by the contractor.

(x) Contractor's or customer's inclination towards a specific type of contract.

(xi) Life span of the project under consideration: For example, if the project in question is to extend over a long period, then it would be subject to variables which may effect the ultimate cost of the project. For example, variation in the price of materials used in the project will certainly effect the overall contracted cost of the project. Therefore, this type of situation calls for a specific type of contract which will take into consideration such future uncertainties.

(xii) Administrative costs.

9.4.2 Types of Contract

This section classifies the types of contracts into classifications I and II. Classification I contracts are classified by method of determining contract cost. On the other hand, classification II contracts are clssified other than by method of determining contract costs. Both these classifications are described below.

CLASSIFICATION I

Various types of contracts fall into this classification [4]. Some of them are as follows.

(i) *Cost reimbursement type contracts:* When the performance costs cannot be estimated with certain accuracy, then cost reimbursement type contracts are used. In this situation the customer pays to the contractor the allowable contract costs. The examples of the cost reimbursement type contracts are cost-plus-fixed-fee and cost-plus-incentive-fee contracts. One of the main disadvantages of these contracts is the administrative burden on both customer and contractor.

(ii) *Letter of Intent:* This is used as a preliminary contractual document before a formal contract is signed. This document authorizes the contractor to start work on the project immediately, with the understanding that the formal contract will be signed at a later stage. According to Reference [5], the formal contract is usually signed within two months.

(iii) *Fixed-Price:* This type of contract is used when the specified price is paid upon the completion of services or delivery of product to the customer. This is the riskiest contract from the contractor standpoint, because there is no provision to take into consideration the increase in cost due to unforseen events in the future. However, this type of contract has the following advantages.

(a) The manufacturer or contractor profit will increase if he is able to cut down the cost of production.

(b) It gives maximum incentive to the contractor to reduce waste.

(c) It gives maximum incentive to the contractors to develop such procurement and production approaches which will save the usage of material and labor.

One should note that these advantages have some overlap.

(iv) *Fixed-Price Incentive Contracts:* This type of contract is used to give incentive to the contractor to improve efficiency and lower costs without lowering the quality of the product in question.

(v) *Fixed-Price with Escalation Clauses:* This type of contract is used when certain contingencies are beyond the control of the contractor or the supplier. Therefore, the escalation clauses are added in this contract. These clauses are concerned with making allowances for future price increases of materials, labor and so on. In other words, if the price of such items increases during the life span of the contract, the contractor or supplier will increase the cost of the contract.

(vi) *Fixed-Price Redeterminable Contracts:* This type of contract is used when sound estimates of labour and materials, accurate specification, and so on are not initially available. Therefore, to reduce the contractor's or supplier's risk, the fixed-price redeterminable contracts are used. At some point during the life span of the contract, the price of the contract is adjusted. This can go either in the upward or in the downward direction.

(vii) *Fixed-Price Contract with Prospective Price Redetermination at a Specified Point in Time:* This contract type is used when the early phase of the project's cost is specified. After the initial phase, at a specified point during the remaining life span of the contract, the contract price is redetermined to go either upward to a specified ceiling or downward. This type of contract has frequent applications in quantity production purchases.

CLASSIFICATION II

As mentioned earlier, only those types of contracts fall into this category which are classified other than by method determining contract costs. Various types of contracts fall into this category. Some of the types of these contracts are listed below [2]:

(i)   Turnkey

(ii) Package

(iii) Serial

(iv) Negotiated

(v)   Continuation

(vi) Competitive

These six types of contracts are described briefly. A turnkey contract is a package deal awarded to the main contractor. This contractor may subcontract specialist works to suitable subcontractors. The major advantage of this type of contract for the customer is that he is relieved from co-ordinating the works of sub-contractors and other details.

The package contract is sometimes known as the package deal. In this contract several related projects are combined and awarded as a single package deal. One should note here that each of these projects could have been awarded as a separate contract.

Another type of contract is known as the serial contract in which the contractor comes to an agreement with the customer to undertake a series of separate contracts over a specified period of time.

The fourth type of contract under classification II is known as the negotiated contract. This is an alternative to the competitive tender contract by which the customer negotiates the contract with a contractor with his liking. This approach is sometimes used when the specifications of the project to be contracted are unclear.

The continuation contract is another type of contract which is used in the contracting business. This type of contract is adopted when a contractor is already doing a good job on a current project (i.e., the project which was awarded to the contractor earlier). On that basis, subject to same terms and conditions the contract of another project is negotiated with the same contractor. Therefore this type of arrangement is called the continuation contract.

The sixth type of contract of this classification is called the competitive contract. This is that type of contract which is negotiated through the formal competitive tendering route, by which a number of tenderers compete for the same project. The final contract is awarded by the customer to the most desirable tenderer.

9.5   SELECTING A CONTRACTOR FOR A PROJECT

Before awarding a contract to a contractor, the customer usually investigates the qualifications of the contractor to ascertain that the project under con-

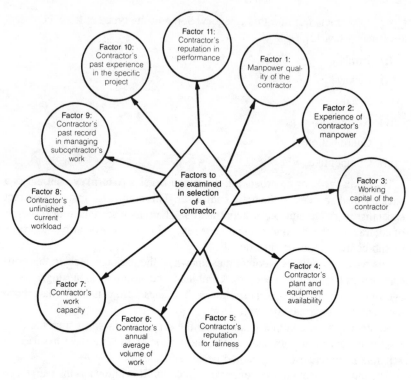

Figure 9.2.
Contractor selection factors.

sideration will be completed successfully. Furthermore, usually a number of contractors with similar qualifications are interested in the project to be contracted. Therefore, to select the most desirable contractor, the customer has to examine the various factors concerning the background qualifications of these contractors. The factors to be examined by the customer when selecting a contractor for the job are given in Figure 9.2. Finally, it is emphasized that it is after examining all these factors that the selection of the contractor is made. Otherwise, the selection of the contractor for the project to be undertaken will not be the most satisfactory one.

9.6 TYPES OF TENDER AND DETERMINING THE PROGRESS OF A CONTRACT

This section briefly describes the types of tenders used in the real life situations and the important purposes of monitoring the contract progress. Both these items are briefly described in the following sections.

9.6.1   Types of Tender

There are various types of tender which are used for contracting. Some of them are as follows:

(i) Selective
(ii) Serial
(iii) Open
(iv) Negotiated

In the case of selective tender, only the select number of tenderers receive the invitation to tender. The second type of tender is known as the serial tender which is similar to the serial contract briefly discussed earlier in the chapter. One may say the serial tender is essentially a standing offer. According to this offer's terms and conditions the contractor agrees to undertake, over a certain period, a series of individual contracts.

The third type of tender is called the open tender, in which case the customer openly advertises the invitation to tender in the press. Lastly, the fourth type of tender is known as the negotiated tender. In this case the customer negotiates with one desirable contractor who makes a tender offer after the settlement of contract and specification details. The tender offer is subject to negotiation.

9.6.2   Determining the Progress of a Contract

To complete the contracted project successfully as planned, after the beginning of the contract the customer must monitor the contractual progress on a regular basis. The same thing applies to the contractor when the same contract has been sub-contracted to others. Both customer and contractor can utilize this contractual progress information to serve various purposes. Examples of such purposes are as follows:

(i) To identify the potential areas of difficulty so that the necessary actions can be planned well in advance
(ii) To provide useful information on already incurred cost for the project under progress. This cost information can be used to perform various types of analysis such as profit, price and so on
(iii) To make interim payments for the work which is already accomplished
(iv) To identify any departure from the project completion date and to obtain information on the expected completion date of the project

9.7   A CONTRACT NEGOTIATION PROCEDURE

In the contracting business, the task of negotiating a contract is very important and has to be systematically accomplished. Essentially, the contract negotiation procedure involves six steps. These steps are shown in Figure 9.3. We should note that the discussion on contract negotiation is presented from the customer's aspect.

The first step of the contract negotiation procedure is solicitation of proposals from desirable contractors. The objective of proposal solicitation is to obtain as many bids as possible from the various contractors to create a competitive environment in which the customer can select the best contractor for the job at the most favorable terms and work conditions.

Figure 9.3.
Steps in a contract negotiation procedure.

Once the proposal solicitation is over, the next step is the evaluation of such proposals. All the proposals are studied from various aspects. In many companies a rating procedure is used to evaluate the different aspects of a proposal. The total ratings of all proposals are compared to provide input to the proposal selection decision.

The third step of the contract negotiation procedure is the selection of a desirable contractor. The factors such as rating, past experience, knowledge of contractors' organizations, and so on are used to select the best contractor for the job.

The next step (the fourth step) of the contract negotiation procedure is one of the most important. This step is concerned with negotiating or bargaining for the best terms and work conditions. In this step a competent bargainer or negotiator will be able to obtain the best terms and working conditions for his firm.

The fifth step of the procedure under discussion is the contract request. This is to request higher management for the actual contract document, once everything else has been settled with the most desirable contractor.

The sixth and final step of the contract negotiation is the awarding of the contract to the most desirable contractor. In this step the appropriate officers of the customer's and the contractor's organizations sign the contract document.

9.8 ATTRIBUTES OF A CONTRACT NEGOTIATOR AND USEFUL KEY POINTS FOR SUCCESSFUL NEGOTIATION

This section presents the qualities of a negotiator and the useful hints for successful contract negotiation. Both these items are described separately as follows.

9.8.1 Attributes of a Negotiator

The success or the failure of awarding a contract depends on how the negotiation is conducted. For example, a contractor or a manufacturer with a good reputation may not be able to secure a contract because of poor qualities of the negotiator. Since securing of a contract depends mainly on how the contract negotiation is conducted, a contract negotiator must possess certain attributes or qualities. Some of them are as follows [6]:

  (i) Ability to effectively express thoughts and arguments
 (ii) Efficient thinking capability
(iii) Facility to think clearly
(iv) Patience and objectivity

(v) Mental ability to determine current situation and take necessary steps

(vi) Ability to translate information such as financial and technical into desirable language

(vii) Impersonality

9.8.2 Useful Key Points for a Successful Negotiation

This section presents a certain number of points which are useful in conducting a successful negotiation. According to Reference [5], the contract negotiator must give consideration to the following points.

(i) *Preparation for Changes:* A good negotiator will always be well prepared for changes in the various aspects of a negotiation. Furthermore, he or she never thinks that any part of the negotiation is fixed or non-negotiable. In addition, a negotiator must also be prepared to take chances in the negotiation.

(ii) *Survey of Work Facilities:* Some customers, before the actual negotiation starts, may like to evaluate the contractor's facilities, its financial history and the concerned manpower. Here the contractor's negotiator must welcome such inspections and be open and frank about the items the potential customer wishes to evaluate.

(iii) *Listening Capacity:* In the contract negotiation the negotiator must be prepared to listen to others instead of talking himself most of the times. All in all, the negotiator must understand the fact that one does not learn the thoughts of others while talking.

(iv) *Treatment of Other Negotiators as Human Beings:* The negotiator must understand that the other negotiators are also human beings. Avoid thinking that the other negotiators are wrong because they are with the other side. In order to speed up the negotiation process, recognition must be given to the following points:

   (a) Development of personal relationships with other negotiators: This could be tried during lunch, coffee breaks and so on.

   (b) Giving credit to the other person's point of view: A good negotiator will give credit to the other man's point of view by assuming that he is probably at least 50 percent correct.

   (c) Treatment of fellow negotiator with respect.

(v) *Alertness of the Negotiator throughout the Contract Negotiation:* To negotiate a contract successfully and efficiently, the negotiator must be alert throughout. Furthermore, a negotiator must take notes of principal points in the ongoing negotiation.

(vi) *Preparation for a Battle of Wits:* During the contract negotiation the opposing negotiator may threaten, shout and so on. Therefore, in order to win demands eventually, the negotiator must not panic, but remain calm and in control.

(vii) *Preparedness for Negotiation:* Whenever a contractor's negotiator is called for a negotiation he must be prepared. Otherwise, chances of winning a contract will be dramatically reduced. Before a negotiator enters a negotiation he must have all the necessary documents and data, e.g., personnel and cost data, details of facilities, specifications, drawings, copy of bid, test data, etc.

9.9 MANAGEMENT OF BIDS

When bidding for a contract, the contractor has to follow a systematic procedure. The three important items concerned with bids are as follows [1]:

(i) Advertisement for invitation to bid

(ii) Information for bidders

(iii) Bid preparation procedure

These three items are described in detail in the following sections.

9.9.1 Advertisement for Invitation to Bid

The objective of advertising is to inform prospective contractors so they can submit a bid for the project in question. This way the necessary competition is introduced among the contractors. Normally, the advertisement for invitation to bid is placed in a newspaper which has necessary circulation to receive enough bids for the project, in order to introduce competitive environments. According to Reference [1], the customer normally allows prospective contractors a minimum of three weeks to submit bids for the project. However, this period may vary from project to project because of size, complexity, etc. The following items should be included in the advertisements:

(i) Names and addresses of the owner (i.e., customer or the party which invites bids) and the person to whom the bids should be submitted

(ii) Project location

(iii) Brief summary of the work to be performed

(iv) Bid and contract sureties

(v) Important items of the projects and their number

(vi) Bids character

(vii) Payment conditions

(viii) Contract award conditions, such as rejection of any or entire bids, lowest responsible bid acceptance reservation, and so on

(ix) Name of the authorized engineer or representatives and statement of their authorization

(x) Plans and specifications information

9.9.2 Information for Bidders

This document is sometimes known as the instructions for bidders document. The document is used to supply all the prospective bidders with the same information on the project in question, so they can submit bids according to the procedure prescribed in the document. The instructions to bidders or information for bidders document contains information similar to that contained in the advertisement. However, this information is more explicit and in more detail. The instructions to bidders document should contain all the items included in the advertisement together with the following:

(i) Starting and finishing time of the project

(ii) Bidder's qualifications

(iii) Responsibility for the accuracy of bidding information, i.e., whether it is of customer (owner) or bidder

(iv) Bid writing and submission approach to be used

(v) Legal implications associated with the awarding of contract or project

(vi) Formalities associated with the bid

(vii) Informalities rejection provision

(viii) Plans and specifications list

(ix) Miscellaneous information (this includes any other information which is not included here)

9.9.3 Bid Preparation Procedure

This sometimes is known as the bid form. The bid form prescribes the approach to be used in the preparation of the bid for the project in question. The bids prepared according to the bid form have many advantages, such as follow:

(i) Simplify the bids comparisons with each other

(ii) Help to detect informalities

(iii) Convenience to bidders

(iv) Help to prevent omissions

(v) Help to insure bid accuracy

According to Reference [1], the bid form should be composed of the following items:

(i) Starting and completion time of the contract

(ii) Information on the sub-contractors to be employed by the bidding company

(iii) Bidding company's qualifications, such as information on available facilities for the project, financial history, and past records

(iv) Bid and contract sureties

(v) Contractor's price for the project in question

(vi) Information on addenda to the specifications and plans which were utilized by the prospective contractor in the preparation of the bid

(vii) Necessary signatures and witnesses

(viii) Declaration that there were no illegal practices (A typical example of such practices is the existence of illegal relations between the customer (owner) representative and the bidder or the prospective contractor.)

(ix) Confirmation that the bidder understood clearly the project plans and specifications

(x) Confirmation (if applicable) that the bidder investigated the project site

9.10 FORMULAS FOR DETERMINING ESCALATION IN PRICE

This section presents two formulas for such purpose. These are as follows.

9.10.1 Formula I

As mentioned earlier in the chapter, there are various types of contracts used to perform contracting work. The price adjustment contracts are one of them. Therefore, this formula is used to determine increase or decrease in contract price due to changes in labor and material costs. Thus, the equation to determine increase or decrease in contract price, C_c, is [2,7]:

$$C_c = C_0 (C_f + C_\ell \cdot \alpha + C_m \beta) - C_o \qquad (9.1)$$

$$\alpha = I_\ell / I_{\ell_t} \qquad (9.2)$$

$$\beta = I_m / I_{mt} \qquad (9.3)$$

where

C_0 is the original price of the contract.

C_f is the non-variable component of the contract price (specified as a fraction of the contract price).

C_ℓ is the fraction of the contract price which varies with labor.

C_m is the fraction of the contract price which varies with materials.
I_ℓ is the labor index applicable at the time of price adjustment.
I_m is the material index applicable at the time of price adjustment.
I_{ℓ_t} is the material index at the base date. In other words, the project contractors use this value to determine the price specified in the tender.
I_{mt} is the material index at the base date. In other words, the project contractors use this value to determine the price specified in the tender.

We should note that in Equation (9.1) the sum of C_f, C_ℓ and C_m must add up to unity.

9.10.2   Formula II

This formula is used to determine the price adjustment factor for labour. According to Reference [7], the formula equation is

$$C_{af} = \alpha_{t1}\left(\frac{A}{R_{b\ell_1}}\right) + \alpha_{t2}\left(\frac{B}{R_{b\ell_2}}\right) \tag{9.4}$$

$$A \equiv R_{c\ell_1} - R_{b\ell_1} \tag{9.5}$$

$$B \equiv R_{c\ell_2} - R_{b\ell_2} \tag{9.6}$$

where

C_{af} is the price adjustment factor for labour.
α_{t1} is the weighting factor for labour force type I. This type of labour force may be representing a specific trade.
α_{t2} is the weighting factor for labour force type II. This type of labour force may be representing general labour force.
$R_{c\ell_1}$ is the labour rate of labour force type I for month in question.
$R_{c\ell_2}$ is the labour rate of labour force type II for the month in question.
$R_{b\ell_1}$ is the base labour rate of labour force type I.
$R_{b\ell_2}$ is the base labour rate of labour force type II.

EXAMPLE 9.1

Assume that the labour force of an organization is classified into two categories, I and II. For both these categories the following data are given:

$$R_{c\ell_1} = 115, \; R_{b\ell_1} = 100, \; R_{c\ell_2} = 70, \; R_{b\ell_2} = 50,$$

$$\alpha_{t1} = 0.6, \; \alpha_{t2} = 0.4$$

Calculate the value of the price adjustment factor for labour.

Thus substituting the specified data into Equations (9.4) through (9.6) results in

$$C_{af} = 0.6 \left(\frac{A}{100} \right) + 0.4 \left(\frac{B}{50} \right)$$

$$A \equiv 115 - 100 = 15$$

$$B \equiv 70 - 50 = 20$$

Thus,

$$C_{af} = 0.6 \left(\frac{15}{100} \right) + 0.4 \left(\frac{20}{50} \right)$$

$$= 0.09 + 0.16 = 0.25 \text{ or } 25\%$$

9.11   SUMMARY

The chapter has briefly explored the various aspects of engineering contracts and their management. These are as follows:

 (i) The specific elements of a contract
 (ii) Essential provisions of a contract
(iii) Contract documents
(iv) Factors for the determination of type of contract
 (v) Types of contracts
(vi) Selection of a contractor
(vii) Types of a tender
(viii) A contract negotiation procedure
(ix) Attributes of a contract negotiator
 (x) Useful hints for successful contract negotiation
(xi) Bidding procedure
(xii) Formulas for determining escalation in price

Each item of the above list is discussed briefly. The first item is concerned with the specific elements of a contract. A court enforceable contract has specific characteristics. The five characteristics of such a contract are presented. Under the second item of the above list, ten essential provisions of a contract are listed.

Under the third item, the contract documents, the nine items comprised in engineering contract documents are briefly described. Items four and five of the list present twelve contract classifications determining factors, and two categories (I and II) of contracts, respectively. Category I includes those contracts which are classified by method of determining contract cost. Seven types of contract are described under this category. On the other hand, category II includes those contracts which are classified other than by method of determining contract costs.

The sixth item of the list presents eleven factors which are useful in selecting a contractor for a project. The seventh item is concerned with the types of tender. The four different types of tender are described.

In the next item (i.e., the eighth item) a six-step procedure to negotiate a contract is described. Items nine and ten present seven qualities of a contract negotiator and seven hints for conducting a successful negotiation, respectively. The eleventh item of the list is concerned with bidding procedure. Three aspects associated with the bidding procedure are described. These are advertisement for invitation to bid, information for bidders, and bid preparation procedure. The last item of the list contains two formulas for determining escalation in price.

Important documents associated with the material presented in the chapter are listed in the reference section.

9.12   EXERCISES

1. Discuss at least five types of contracts with respect to pricing.
2. What are the essential items of a contract?
3. What are the essential elements of a contract which make it enforceable through court?
4. Discuss the determining factors for types of contracts.
5. What are the factors which are to be examined when selecting a contractor?
6. Describe at least nine qualities of a contract negotiator.
7. What are the principal items which are to be included in the instructions to bidders?
8. Describe briefly the following items:
   (i) Open tender
   (ii) Selective tender
   (iii) Negotiated tender
   (iv) Competitive contract
   (v) Turnkey contract
9. Describe the six-step procedure used in negotiating a contract.

9.13 REFERENCES

1. Abbett, R. W. *Engineering Conracts and Specifications.* New York:John Wiley & Sons (1963).

2. Horgan, M. O. C. and F. R. Roulston. *The Elements of Engineering Contracts.* Woodcote Grove, Epsom, Surrey, England:W. S. Atkins & Partners (August 1977).

3. Hajek, V. G. *Management of Engineering Projects.* New York:McGraw-Hill Book Company (1977).

4. Hayes, G. E. and H. G. Romig. "Modern Quality Control," Encino, CA:BRUCE, A Division of Benziger Bruce & Glencoe, Inc. (1977).

5. Hicks, T. G. *Successful Engineering Management.* New York:McGraw-Hill Book Company (1966).

6. Dhillon, B. S. *System Reliability, Maintainability and Management.* Princeton, NJ:Petrocelli Book Inc. (1983).

7. Marsh, P. D. V. *Contracting for Engineering and Construction Projects.* Aldershot, England:Gower Publishing Company Limited (1981).

CHAPTER 10

Techniques for Making Better Engineering Management Decisions

10.1 INTRODUCTION

Since World War II there has been a significant growth in quantitative management techniques. Furthermore, the increase in the use of computers and the modern problem complexities have enhanced the importance of many of these techniques. Linear and non-linear programming techniques are the prime examples. Today, techniques such as linear programming, decision trees, exponential smoothing and discounted cash flow analysis are used to solve management associated problems.

Therefore, this chapter presents various techniques which directly or indirectly provide useful inputs to make sound management decisions.

10.2 OPTIMIZATION TECHNIQUES

Two widely known techniques for optimization are briefly explained in this section. These are the Lagrangian multiplier and linear programming. The Langrangian multiplier, sometimes known as "the technique of undetermined multipliers," is due to Joseph Lagrange, the French mathematician, who discovered it in the eighteenth century. This method allows the optimization of functions subject to defined constraints without the elimination of any of the concerned variables.

The history of linear programming goes back to 1947 when George B. Dantzig developed the simplex method. With the discovery of the concept of duality by J. Von Neumann in the same year, the simplex technique received even wider acceptance. Ever since then, several researchers have contributed to the subject of linear programming. It is the simplest and the most widely used technique for the purpose of optimization subject to constraints.

10.2.1 Lagrangian Multiplier

In this section, this technique is demonstrated for a two-variable function [1]. However, on similar lines it can be extended for n variables.

Assume that we have defined function $f(y_1,y_2)$ subject to the constraint function $k(y_1,y_2) = 0$. With the aid of these two functions the Lagrange function, $L(y_1,y_2,\lambda)$ is formulated as follows:

$$L(y_1,y_2,\lambda) = f(y_1,y_2) + \lambda\, k(y_1,y_2) \tag{10.1}$$

subject to the following necessary conditions for estimating a relative maximum or minimum value:

$$\frac{\partial L(y_1,y_2,\lambda)}{\partial y_1} = 0 \tag{10.2}$$

$$\frac{\partial L(y_1,y_2,\lambda)}{\partial y_2} = 0 \tag{10.3}$$

$$\frac{\partial L(y_1,y_2,\lambda)}{\partial \lambda} = 0 \tag{10.4}$$

Expressions for y_1,y_2 and λ can be obtained by solving the simultaneous Equations (10.2)–(10.4). For more information on this technique consult Reference [1].

EXAMPLE 10.1

Find the critical point for $f(y_1,y_2) = y_1^2 + y_2^2$, subject to constraint $k(y_1,y_2) = 2y_1 - y_2 - 4 = 0$.

Thus the Lagrange function is formulated as follows:

$$L(y_1,y_2,\lambda) = y_1^2 + y_2^2 + \lambda(2y_1 - y_2 - 4) \tag{10.5}$$

Taking the partial derivatives of Equation (10.5), with respect to y_1,y_2 and λ, results in

$$\frac{\partial L(y_1,y_2,\lambda)}{\partial y_1} = 2y_1 + 2\lambda \tag{10.6}$$

$$\frac{\partial L(y_1,y_2,\lambda)}{\partial y_2} = 2y_2 - \lambda \tag{10.7}$$

$$\frac{\partial L(y_1,y_2,\lambda)}{\partial \lambda} = 2y_1 - y_2 - 4 \tag{10.8}$$

Setting the left-hand side of Equations (10.6)–(10.8) equal to zero leads to

$$2y_1 + 2\lambda = 0 \tag{10.9}$$

$$2y_2 - \lambda = 0 \qquad (10.10)$$

$$2y_1 - y_2 - 4 = 0 \qquad (10.11)$$

Solving Equations (10.9)–(10.11) yields

$$y_1 = 8/5$$

$$y_2 = -8/10$$

$$\lambda = -8/5$$

Thus the critical point of f, subject to specified condition, is $(8/5, -8/10)$.

10.2.2 Linear Programming

In real life situations the objective and constraint functions will be rather complex [2]. However, the simplest form of linear programming problem formulation may be expressed as follows:

maximize (or minimize) $Y = a_1z_1 + a_2z_2 + a_3z_3 + --- + a_kz_k$ \qquad (10.12)

subject to

$$b_{11}z_1 + b_{12}z_2 + b_{13}z_3 + --- + b_{1i}z_i + --- + b_{1k}z_k \ (\leq,=,\geq)P_1$$
$$(10.13)$$

$$b_{21}z_1 + b_{22}z_2 + b_{23}z_3 + --- + b_{2i}z_i + --- + b_{2k}z_k \ (\leq,=,\geq)P_2$$
$$(10.14)$$

$$b_{j1}z_1 + b_{j2}z_2 + b_{j3}z_3 + --- + b_{ji}z_i + --- + b_{jk}z_k \ (\leq,=,\geq)P_j$$
$$(10.15)$$

$$b_{n1}z_1 + b_{n2}z_2 + b_{n3}z_3 + --- + b_{ni}z_i + --- + b_{nk}z_k \ (\leq,=,\geq)P_n$$
$$(10.16)$$

$$z_1 \geq 0$$

$$z_2 \geq 0 \qquad (10.17)$$

$$z_k \geq 0$$

The symbols associated with Equations (10.12) through (10.16) are defined below:

k denotes the total number of variables.

n denotes the total number of constraints.

a_i denotes the ith constant for $i = 1, 2, ---, k$.

b_{ji} denotes the (j,i) constant for $j = 1, 2, ---, n$ and $i = 1, 2, ---, k$.

P_j denotes the jth constant or resource, for $i = 1, 2, ---, n$.

z_i denotes the ith variable, for $i = 1, 2, ---, k$.

Y denotes the total profit or the total cost.

Equation (10.12) is known as the objective function, the profit function, or the cost function. Therefore, the idea in this formulation is to either maximize or minimize Equation (10.12) subject to constraints (10.13)–(10.17).

EXAMPLE 10.2

For Equations (10.12)–(10.17) the following values for the symbols are defined:

$$n = 2, k = 3, a_1 = 5, a_2 = 10, a_3 = 20$$

$$b_{11} = 4, b_{12} = 6, b_{13} = 5, P_1 \leq 100$$

$$b_{21} = 1, b_{22} = 5, b_{23} = 3, P_2 \leq 200$$

Write down the resulting equations by assuming that the objective function is to be maximized.

Thus substituting the specified data in Equations (10.12)–(10.17) leads to

$$\text{maximize } Y = 5z_1 + 10z_2 + 20z_3$$

subject to

$$4z_1 + 6z_2 + 5z_3 \leq 100$$

$$z_1 + 5z_2 + 3z_3 \leq 200$$

$$z_1 \geq 0$$

$$z_2 \geq 0$$

$$z_3 \geq 0$$

Generally, in order to find solutions to the problems which can be expressed by Equations (10.12)–(10.17), the simplex method is used. However, if a problem has only two variables, it is usually more efficient to solve it graphically. The simplex method is described in detail in Reference [2]. However, this section presents only the graphical approach. The approach is demonstrated in Example 10.3.

EXAMPLE 10.3

An Engineering manufacturer produces two types, α and β, of a product. The product uses three different components, A, B and C. The company has 20 components of type A, 30 components of type B and 10 components of type C.

The product type α uses two components of type A and four components of type B. However, the product type β uses five components of type C and two components of type B.

Each unit of α and β makes a profit of 10 and 6 dollars, respectively. Determine the quantities of α and β to be produced to maximize the profit. Calculate the total maximum profit.

Assuming that we make z_1 units of product type α and z_2 units of product type β, the total profit Y is given by

$$Y = 10z_1 + 6z_2 \tag{10.18}$$

In this problem, we use

- $2z_1$ of type A components
- $5z_2$ of type C components
- $4z_1 + 2z_2$ of type B components

With the aid of specified data and the above formulations, the resulting linear programming problem is formulated as follows:

$$\text{Maximize } Y = 10z_1 + 6z_2 \tag{10.19}$$

subject to

$$2z_1 \leq 20 \tag{10.20}$$

$$5z_2 \leq 10 \tag{10.21}$$

$$4z_1 + 2z_2 \leq 30 \tag{10.22}$$

and

$$z_1 \geq 0 \tag{10.23}$$

$$z_2 \geq 0 \tag{10.24}$$

The plots of Equations (10.19) through (10.24) are shown in Figure 10.1. As shown in Figure 10.1, the company profit will be optimum at point F. At this point all the constraints are satisfied and the profit is maximum. This can be observed from two profit lines shown in Figure 10.1. In this plot, F is the most extreme point through which the profit line can pass and still satisfy all the constraints.

Thus, at point F the optimum values of z_1 and z_2 are 6.5 and 2, respectively. Substituting those values into Equation (10.19) yields

$$Y = 10(6.5) + 6(2)$$

$$= 77$$

The total maximum profit will be 77 dollars.

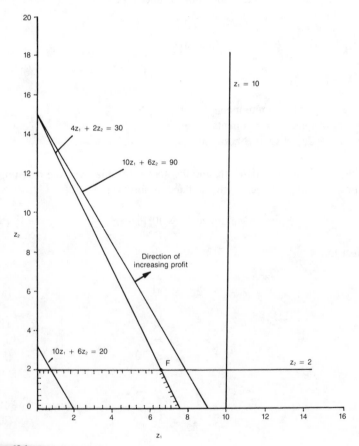

Figure 10.1.
The graphical plot.

10.3 DISCOUNTED CASH FLOW ANALYSIS

Every one of us is directly or indirectly exposed to interest transactions. In the engineering industry, money borrowing and lending transactions are being carried out every day. Furthermore, in various engineering investment decisions, the time value of money plays an important role. Therefore it is essential for the engineers to have some knowledge of engineering economics. The basics of engineering economics are presented in this section.

10.3.1 Simple Interest

This is that interest which is calcaulated on the original sum of money, called the original principal, for the period in which the lent or borrowed sum is being utilized. The simple interest, S_I, is given by

$$S_I = (M)\,(i)\,(k) \tag{10.25}$$

where

M is the principal amount lent or borrowed.
i is the interest rate per period (this is normally a year).
k is the interest periods (these are usually years).

The total amount of money, M_t, after the specified lent or borrowed period is given by

$$M_t = M + S_I$$
$$= M + (M)\,(i)\,(k) = M[1 + (i)\,(k)] \tag{10.26}$$

EXAMPLE 10.4

A person borrowed $2,000 at a simple interest rate of 10% per year. The total amount of money is to be paid back at the end of five years. Determine the amount of money due at the end of that specified period.

The simple interest, S_I, for the five-year period from Equation (10.25) is given by

$$S_I = (M)\,(i)\,(k) = (2,000)\,(0.1)\,(5) = \$1,000$$

Thus the total amount due at the end of the five-year period from Equation (10.26) is

$$M_t = M + S_I = 2,000 + 1,000 = 3,000 \text{ dollars}$$

10.3.2 Compound Interest

In this case, at the end of each equal specified period, the earned interest is added to the original principal or amount lent or borrowed at the beginning of that period. Thus, this new principal, or amount, acts as a principal for the next period and the process continues. This way the interest is compounded into principal. To calculate compound amount, M_{ck}, the resulting formula is developed as follows.

At the end of the first period (or year) the amount

$$M_{c1} = M + M(i) = M(1 + i) \qquad (10.27)$$

where

M is the original principal or money lent or borrowed.
i is the interest rate per period (or, normally, per year).

At the end of the second period (or year), the amount

$$M_{c2} = [M(1 + i)] \cdot (1 + i) \qquad (10.28)$$

where $M(1 + i)$ is the principal for the second period.

Similarly, at the end of the third period (or year), the amount

$$M_{c3} = [M (1 + i)^2] \cdot (1 + i) \qquad (10.29)$$

where $M(1 + i)^2$ is the principal for the third period.

At the end of kth period (or year), the amount

$$M_{ck} = [M(1 + i)^{k-1}] \cdot (1 + i) \qquad (10.30)$$

$$= M(1 + i)^k$$

The total compound interest, C_{Ik}, earned after the kth period or year, is given by

$$C_{Ik} = M_{ck} - M \qquad (10.31)$$

EXAMPLE 10.5

An engineering organization borrowed $70,000 at a compound interest rate of 15% per annum. The borrowed sum has to be paid back at the end of three years. Calculate the compound interest after three years.

In the example, the data is specified for the following items:

$$M = 70{,}000, \; i = 15\%, \; k = 3$$

Utilizing the above data in Equation (10.30) yields

$$M_{c3} = 70{,}000 \, (1 + 0.15)^3 = \$106{,}461.25$$

Thus substituting the above result and the specified data for M into Equation (10.31) yields

$$c_{I3} = M_{c3} - M = 106{,}461.25 - 70{,}000$$

$$= \$36{,}461.25$$

The compound interest after three years will be \$36,461.25.

10.3.3 Present Worth

The present worth of a single payment by rearranging Equation (10.30) is given by

$$M = M_{ck}(1 + i)^{-k} \tag{10.32}$$

In simple terms, the above formula is used to obtain the present worth, M, of money, M_{ck}, after k periods, discounted at the periodic interest rate of i. Sometimes Equation (10.32) is written in the following form:

$$M = M_{ck} \, e^{-[\ln(1+i)] \, k} \tag{10.33}$$

EXAMPLE 10.6

An engineering company has purchased a new machine tool. After ten years' usage its estimated salvage value is \$10,000. The estimated value of the annual interest rate compounded annually is 10%. Compute the present worth of the salvage value.

In this example, the data is specified for the following items used in Equation (10.32):

$$k = 10, \; M_{c10} = \$10{,}000, \text{ and } i = 10\%$$

Thus substituting the above specified data into Equation (10.32) results in

$$M = 10,000 \ (1 + 0.1)^{-10} = \$3,855.43$$

The present worth of the machine tool salvage value is \$3,855.43.

10.3.4 Formula for Uniform Periodic Payments

The detailed derivation of this formula is given in References [3,5]. In the development of the formula, it is assummed that at the end of each of the k periods or years, the depositor adds D amount of money. The money is invested at the interest rate i, compounded annually or periodically. Thus the total amount of money

$$M_{ck} = D(1 + i)^{k-1} + D(1 + i)^{k-2} + --- + D(1 + i) + D$$

$$(10.34)$$

where

$D(1 + i)^{k-1}$ denotes the value of D after $(k - 1)$ periods or years.
$D(1 + i)^{k-2}$ denotes the value of D after $(k - 2)$ periods or years.
$D(1 + i)$ denotes the value of D after one period or year.
D denotes the money which has just been deposited.

It should be noted in Equation (10.34) that the D amount of money is first time deposited at the end of the first year or period.

EXAMPLE 10.7

A person deposits \$2,000 at the end of each year for the next three years at the annual interest rate of 15% compounded annually. Calculate the total sum of money after the three-year period.

At the end of the first year, due to first deposit, the total amount of money is

$$M_{c1} = 2,000$$

At the end of the second year, after the second deposit, the total amount of money is

$$M_{c2} = 2,000 \ (1 + 0.15) + 2,000$$

Similarly, at the end of the third year, after the third or last deposit, the total amount of money is

$$M_{c3} = [2,000(1 + 0.15) + 2,000] \ (1 + 0.15) + 2,000 \qquad (10.35)$$

Rearranging the right-hand side of the above equation leads to

$$M_{c3} = (2{,}000)(1 + 0.15)^2 + (2{,}000)(1 + 0.15) + (2{,}000) \qquad (10.36)$$

It is to be noted here that for $k = 3$, $D = \$2{,}000$, and $i = 0.15$, Equation (10.34) corresponds to the above resulting equation. Thus from Equation (10.36), the total amount of money after the three-year period is

$$M_{c3} = \$6{,}945$$

Rewriting Equation (10.34) yields

$$M_{ck} = D[(1 + i)^{k-1} + (1 + i)^{k-2} + \ldots + (1 + i) + 1] \qquad (10.37)$$

One can easily see that this is a geometric series. To find the sum of this series we multiply both sides of Equation (10.34) by $(1 + i)$ as follows:

$$M_{ck}(1 + i) = D[(1 + i)^1 + (1 + i)^2 + \ldots + (1 + i)^k] \qquad (10.38)$$

Subtracting Equation (10.37) from Equation (10.38) yields

$$M_{ck}(1 + i) - M_{ck} = D[(1 + i)^k - 1] \qquad (10.39)$$

Rewriting Equation (10.39) leads to

$$M_{ck} = (i)^{-1}[(1 + i)^k - 1]D \qquad (10.40)$$

EXAMPLE 10.8

By utilizing the given data in Example 10.7, calculate the total amount of money with the aid of Equation (10.40).

In Example 10.7, the data is specified for the following items:

$$k = 3, D = \$2{,}000, i = 0.15$$

Thus substituting the above data into Equation (10.40) results in

$$M_{c3} = (0.15)^{-1}[(1 + 0.15)^3 - 1](2{,}000) = \$6{,}945$$

The above result is the same as obtained in Example 10.7.

10.3.5 Present Value of Uniform Periodic Payments

In this situation, we are concerned with the same problem as that of the previous section (i.e., 10.3.4). However, in this situation the only difference is that we wish to find the present value of uniform periodic payments after k

periods or years, instead of the total amount. Thus, the present worth, *PW*, of uniform periodic payments is given by

$$PW = D[(1 + i)^{-1} + (1 + i)^{-2} + --- + (1 + i)^{-k}] \qquad (10.41)$$

Again, this is a geometric series. The sum of this series can be found the same way as that of Equation (10.37) with one exception: multiply both sides of Equation (10.41) with $(1 + i)^{-1}$. Thus, the resulting present-worth formula is

$$PW = D(i)^{-1} [1 - (1 + i)^{-k}] \qquad (10.42)$$

EXAMPLE 10.9

A company has purchased a machine for $50,000. The useful life of the machine is 10 years and its salvage value is zero.

At the end of each useful year the earning from the machine utilization is $8,000. This sum of money is invested at the annual interest rate of 8% compounded annually. Determine if the purchase of the machine is going to be a profitable venture.

In this example the data for the elements of Equation (10.42) are as follows:

$$D = \$8,000, k = 10 \text{ years}, i = 0.08$$

Thus substituting the above data into Equation (10.42) yields the present worth of the total earnings:

$$PW = (8,000)(0.08)^{-1} [1 - (1 + 0.08)^{-10}]$$

$$= \$53,680.65$$

Thus the net profit, *NP*, from the machine will be

NP = (total earnings from the machine) − (machine procurement cost)

$$= \$53,680.65 - \$50,000$$

$$= \$3,680.65$$

The decision to purchase the machine will be a profitable investment.

10.4   DEPRECIATION TECHNIQUES

The term depreciation means a decline in value. It is understood that as the engineering products become older, they become less valuable. At the end of their useful life they are retired or replaced with new ones. In order to take into consideration the change in the value of the product, the depreciation charges are made during the useful life of the engineering products. Therefore, this section presents the following three depreciation techniques [6].

10.4.1   Declining-Balance Depreciation Method

This is a frequently used method of depreciation. The method dictates the accelerated write-off of the product worth in its early productive years and corresponding lower write-off near the end of the useful life years. The depreciation rate, α_d, is given by

$$\alpha_d = 1 - A^{1/m} \tag{10.43}$$

where

$$A \equiv s/c \tag{10.44}$$

where

c denotes the procurement cost of the product.
s denotes the salvage value of the product.
m denotes the product useful life in years.

The declining balance technique requires a positive value for s. The product book value, $V_{bv}(M)$, at the end of year M is given by

$$V_{bv}(M) = c[1 - \text{depreciation rate}]^M$$

$$= c[1 - \alpha_d]^M \tag{10.45}$$

$$= c \cdot A^{(M/m)}$$

The annual depreciation charge, $DC(M)$, at the end of year M is given by

$$DC(M) = V_{bv}(M - 1) \left[1 - \left(\frac{s}{c} \right)^{1/m} \right] \tag{10.46}$$

EXAMPLE 10.10

An engineering product procurement cost is $40,000 and its useful life is 8 years. The salvage value of the product is $2,000. Determine the product book value at the end of the four-year period by utilizing the declining-balance method.

In this example the data is specified for the following items:

$$c = \$40,000, \; s = \$2,000, \; m = 8 \text{ years}, \; M = 4 \text{ years}$$

Substituting the above data into Equation (10.45) leads to

$$V_{bv}(4) = (40,000) \left(\frac{2,000}{40,000} \right)^{4/8}$$

$$= \$8,944.27$$

10.4.2 Straight-Line Depreciation Method

This is the most widely used technique and is very easy to apply. In this method it is assumed that the annual depreciation is constant during the productive life of the product. Thus the annual depreciation charge is given by

$$DC_a = (c - s)/m \tag{10.47}$$

The book value of the product at the end of year M is

$$V_{bv}(M) = c - (c - s)\beta \tag{10.48}$$

where

$$\beta \equiv M/m \tag{10.49}$$

EXAMPLE 10.11

An electrical system has been recently procured by a company at the cost of $200,000. Its estimated useful life period is 7 years. At the end of the product life period, the system salvage value will be $30,000. Assume that the system annual depreciation is constant. Determine the electrical system annual depreciation charge.

Substituting the specified data into Equation (10.47) results in

$$DC_a = (200{,}000 - 30{,}000)/7$$

$$= \$24{,}285.714$$

10.4.3   Sum-of-Digits Depreciation Method

The name of the technique is obtained from the calculation approach. For this technique, in the initial years the depreciation charge is larger than in the final years of the product useful life. The annual depreciation charge, DC_a, is given by

$$DC_a = \frac{\text{Remaining years of product life}}{\text{Sum of the digits for the total life}}(c - s)$$

$$(10.50)$$

$$= \frac{(m - M + 1)}{\displaystyle\sum_{i=1}^{m} i}(c - s)$$

The sum of the denominator of Equation (10.50) is given by

$$\sum_{i=1}^{m} i = \frac{m(m + 1)}{2} \qquad (10.51)$$

Thus substituting Equation (10.51) into Equation (10.50) leads to

$$DC_a(M) = \frac{2(m - M + 1)}{m(m + 1)} \cdot (c - s) \qquad (10.52)$$

The product book value at the end of year M is

$$V_{bv} = \frac{2(c - s)\,[1 + 2 + \text{---} + (m - M)]}{m(m + 1)} + s \qquad (10.53)$$

EXAMPLE 10.12

A company has spent \$50,000 to purchase a mechanical device. The expected useful life of the device is 10 years. At the end of the useful life period the salvage value of the device will be \$4,000. Determine the annual depreciation charge during the second year of use.

Thus substituting the specified data into Equation (10.52) results in

$$DC_a(2) = \frac{2(10 - 2 + 1)}{10(10 + 1)} (50,000 - 4,000)$$

$$= \frac{18}{110} \cdot (46,000)$$

$$= \$7,527.273$$

10.5   BUSINESS OPERATIONS ANALYSIS

This section presents a mathematical model for performing analysis of business operations. This model is concerned with finding the point of maximum profit, the point of maximum investment rate and the point of maximum economic production when the cost of production is described by a parabolic equation [7]. Thus we assume that the cost of production is defined by

$$c = \alpha z^2 + \beta z + \theta \tag{10.54}$$

where

c is the cost of production.
z is the number of units produced.
α is a constant.
β is a constant.
θ is a constant.

Differentiating Equation (10.54) with respect to z yields the following expression for the incremental cost:

$$\frac{dc}{dz} = 2z\,\alpha + \beta \tag{10.55}$$

The unit cost, c_u, is given by

$$c_u = \frac{c}{z} = (\alpha z^2 + \beta z + \theta)/z$$

(10.56)

$$= \alpha z + \beta + \frac{\theta}{z}$$

By equating the right-hand side of Equation (10.55) to the unit selling price, p_s, we get

$$p_s = 2z\,\alpha + \beta$$

(10.57)

Thus, from Equation (10.57), the quantity of units to be produced to maximize profit is given by

$$z = (p_s - \beta)/2\alpha$$

(10.58)

Equating the right-hand side of Equation (10.55) to the right-hand side of Equation (10.56) yields

$$2z\alpha + \beta = \alpha z + \beta + \frac{\theta}{z}$$

(10.59)

Solving Equation (10.59) for z yields the following expression for the quantity of units to be produced to maximize investment rate:

$$z = \left(\frac{\theta}{\alpha}\right)^{1/2}$$

(10.60)

The investment rate, I_r, is defined by

$$I_r = \frac{\text{profit}}{\text{investment}}$$

(10.61)

From Reference [7], to find the point of maximum economic production, we have

$$\frac{dc}{dz} = \frac{p_s}{1 + r}$$

(10.62)

where r denotes the lowest acceptable rate of return.

Equating the right-hand side of Equation (10.62) to the right-hand side of Equation (10.55) yields

$$\frac{p_s}{1 + r} = 2z\alpha + \beta \tag{10.63}$$

Solving Equation (10.62), the maximum economic production quantity is given by

$$z = \frac{1}{2\alpha}\left(\frac{p_s}{1 + r} - \beta\right) \tag{10.64}$$

EXAMPLE 10.13

A company manufactures a certain type of engineering product. Each unit of the product selling price is $400. The cost of production is described by

$$c = 0.4z^2 + 200z + 15,000 \tag{10.65}$$

where z denotes the number of units produced.

Determine the maximum profit.

In this example, the data is specified for the following items:

$$\alpha = 0.4, \ \beta = 200, \ \theta = 15,000, \ p_s = \$400$$

Substituting the above data into Equation (10.58) yields the following quantity of units to be produced to maximize profit:

$$z = (400 - 200)/2(0.4) = 250 \text{ units}$$

Hence,

$$\text{total sales revenue} = 250 \times 400 = \$100,000$$

The cost to manufacture 250 units from Equation (10.65) is

$$c = 0.4(250)^2 + (200)\,(250) + 15,000$$

$$= \$90,000$$

Thus,

maximum profit = (total sales revenue) − (cost to manufacture 250 units)

$$= \$100,000 - \$90,000$$

$$= \$10,000$$

10.6 FORECASTING

An engineering manager makes many decisions which deal with some point in the future. Therefore, in order to make such decisions intelligently, a projection into the future is necessary. For example, when planning for the development of a new product, it would not be wise to plan without having a knowledge of the new product demand in the future. In order to predict future demand for a product, forecasting criteria are usually followed. Forecasting is the process of estimating a future event by utilizing the past data.

When using forecasted results, care must be given to the fact that these forecasts will be valid only if they are used according to specified conditions or the conditions under which the forecasts are made. For example, forecasts made by assuming certain economic, governmental, market, and other factors will be realistic only if they are used subject to those conditions.

10.6.1 Forecasting Techniques

There are various techniques which are utilized to make product demand and other forecasts. However, in forecasting there is essentially no procedure for making absolutely sure that a certain technique is most useful for a particular forecast. The forecasting technique selection may be subject to any one or more of the following factors [9]:

(i) Data accuracy

(ii) Forecast development cost

(iii) Prediction interval length

(iv) Expected accuracy from the forecasted result

(v) Past data availability

(vi) Complexity of factors affecting operations in time to come

(vii) Analysis time

According to Reference [8], forecasting techniques can be categorized into the following two groups:

 (i) empirical techniques

 (ii) analytical techniques

Both these groups of techniques are discussed separately in the following sections.

EMPIRICAL TECHNIQUES

These techniques use input data coming from subjective judgements. There are three commonly used empirical forecasting techniques. These are as follows.

  (i) *Jury of Executive Opinion Approach:* This approach utilizes forecasts made by the various executives of a company in question. The consensus reached from this data is usually reviewed by a group of senior executives or by the company president. In addition, this approach is normally used, if the cost is low, as a first step in the forecasting procedure. However, the method suffers from the fact that it is heavily dependent on subjective opinions without the support of any actual data.

 (ii) *Sales Force Opinion Approach:* This technique is very similar to the jury of executive opinion approach. However, the only major difference is that in this approach the opinions of sales force personnel are used instead of those of executive personnel.

(iii) *Customer's Expectation Technique:* The input data for this technique is obtained from the users of the company product or service. This information is used to develop the forecast. Care must be given when developing a program for soliciting the users to obtain desired information. According to Reference [8], this technique is very practical when:

   (a) the number of users is small.

   (b) a small number of big companies dominates the demand for the service or goods.

   (c) the company needing forecasting information is small in size and the resources required to use other forecasting approaches are not within reach.

ANALYTICAL TECHNIQUES

These are those techniques which make use of mathematics to obtain desired results. Some of the useful mathematical models used for forecasting are given in Figure 10.2.

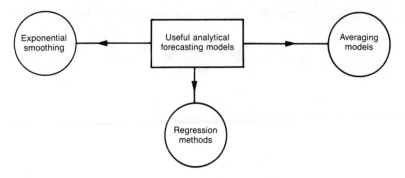

Figure 10.2.
Some of the useful analytical techniques used for forecasting.

In this section, only averaging models and exponential smoothing are presented, simply to familiarize the reader with the subject of forecasting. However, for further reading on the subject, one should consult References [8]–[10]. Averaging and exponential smoothing models are described below.

MODEL I: SIMPLE AVERAGE

In this model, the demands of all earlier periods are given equal weight. Thus the simple average, A_s, of past data is given by

$$A_s = \sum_{i=1}^{k} d_i / k \tag{10.66}$$

where

d_i denotes the demand for the *i*th past period (in other words, the demand that occurred *i* periods earlier).
k denotes the total number of all the past demand periods.

The calculated value of A_s will be the forecasted value of demand for all future periods. The advantage of this model is that the effects of randomness are minimized because all the past periods' data are used in the calculation.

EXAMPLE 10.14

In the past five four-month periods, the demand for a certain company product has been 100, 120, 100, 110 and 90 units, respectively. Calculate the simple average. Substituting the given data in Equation (10.66) yields

$$A_s = \frac{d_1 + d_2 + d_3 + d_4 + d_5}{k}$$

$$= \frac{100 + 120 + 100 + 110 + 90}{5}$$

$$= 104$$

Thus the forecast for each of the future four-month periods is 104 units.

MODEL II: SIMPLE MOVING AVERAGE

This model is similar to the earlier model. However, in this model the data from several recent past periods are averaged and used as a forecast for the next period. The number of recent past-periods data to be used in computing the moving average has to be decided by the forecaster and then held constant throughout. In this situation, the average "moves" over time because the oldest period demand data is discarded and the newest period demand data is included in the calculation of forecast for the next period. Thus the process continues. This process is shown in Figure 10.3 for the fixed past periods. In other words, Figure 10.3 shows that the data for only four recent past periods are used to compute moving average. For example, in Figure 10.3 the data for periods 1, 2, 3, and 4 are used to calculate moving average results for the forecast period 1. Similarly, the data for periods 2, 3, 4 and forecast period 1 are used to calculate moving average results for the forecast period 2. In this situa-

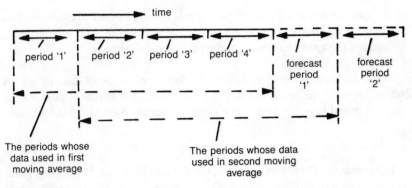

Figure 10.3.
Description of a four-period moving average.

tion, the period 1 data is discarded and the forecast period 1 data added. Thus the g-period moving average A_m is given by

$$A_m = \frac{\text{Total of demands for past } g\text{-periods}}{\text{Periods used to calculate moving average}}$$

(10.67)

$$= \left(\sum_{i=1}^{g} d_i \right) \Big/ g$$

where d_i denotes the ith period demand; for $i = 1$ means the oldest period demand in the g-period average, for $i = g$ means the most recent period demand in the g-period average.

EXAMPLE 10.15

The monthly sales for a certain company product from August to December was 60, 80, 70, 70 and 60 units, respectively. Calculate the five-month moving average and the forecast for the month of January. Thus substituting the specified data into Equation (10.67) yields

$$A_m = \left(\sum_{i=1}^{5} d_i \right) /5 = \frac{60 + 80 + 70 + 70 + 60}{5} = 68 \text{ units}$$

Thus the January forecast is 68 units.

MODEL III: WEIGHTED MOVING AVERAGE

This model is similar to Model II. However, in this model the forecaster assigns a weight to each demand of past periods in question. This way, the forecaster can assign weights to past g-period demands according to his desire instead of having equal weight for all the past period demands. The weighted moving average, A_{wm}, is given by

$$A_{wm} = w_1 d_1 + w_2 d_2 + \text{---} + w_g d_g$$

(10.68)

$$= \sum_{i=1}^{g} w_i d_i$$

where

$$\sum_{i=1}^{g} w_i = 1.0 \qquad (10.69)$$

$$0 \leq w_i \leq 1 \qquad (10.70)$$

and w_i denotes the weight associated with ith period demand d_i; for $i = 1$ means the weight associated with the oldest period demand, for $i = g$ means the weight associated with the most recent period demand.

EXAMPLE 10.16

An engineering company during the months of April, May, June, July and August sold 80, 70, 50, 60 and 70 units of an engineering product, respectively. The demands for the recent periods of July and August are weighted heavier than those of April, May and June periods. Thus, the weights for the demands of periods April, May, June, July and August are 0.1, 0.1, 0.1, 0.3 and 0.4, respectively. By utilizing Equation (10.68), forecast the demand for the period of September.

In this example the data is specified for the following items of Equation (10.68):

$$d_1 = 80, \; d_2 = 70, \; d_3 = 50, \; d_4 = 60, \; d_5 = 70$$

$$w_1 = 0.1, \; w_2 = 0.1, \; w_3 = 0.1, \; w_4 = 0.3, \; w_5 = 0.4$$

$$g = 5$$

Thus substituting the above specified data into Equation (10.68) results in

$$A_{wm} = (0.1)(80) + (0.1)(70) + (0.1)(50) + (0.3)(60) + (0.4)(70)$$

$$= 66 \text{ units}$$

The demand forecast for the month of September is 66 units.

MODEL IV: EXPONENTIAL SMOOTHING

This model is widely known and is frequently utilized in operations management. The reasons for its wide usage may be as follows [10]:

(i) Availability in the standard computer software packages
(ii) The requirements for data storage and computation facilities are relatively low.

Basically, exponential smoothing is an averaging method and is useful for forecasting one period ahead. In this approach, the most recent past period demand is weighted most heavily. In a continuing manner the weights assigned to successively past period demands decrease according to exponential law. This is the reason that the technique is named exponential smoothing. For this approach the forecast, f_t, for demand one period ahead is given by

$$f_t = \theta \begin{bmatrix} \text{Actual demand for the most} \\ \text{recent past period} \end{bmatrix} + (1 - \theta) \begin{bmatrix} \text{Demand forecast for the} \\ \text{most recent past period} \end{bmatrix}$$

$$f_t = \theta d_{t-1} + (1 - \theta)f_{t-1}, \text{ for } 0 \leq \theta \leq 1 \tag{10.71}$$

where

t is the time period.
f_t is the forecast for demand one period ahead.
d_{t-1} is the actual demand for the most recent past period.
f_{t-1} is the demand forecast for the most recent past period.
θ is the weighting factor or the smoothing constant.

Figure 10.4 shows various equal time periods.
From Equation (10.71), the forecast for the period just ending is given by

$$f_{t-1} = \theta \, d_{t-2} + (1 - \theta) \, f_{t-2} \tag{10.72}$$

where

f_{t-2} is the forecast for the time period $(t-2)$.
d_{t-2} is the actual demand for the time period $(t-2)$.

Similarly, the forecast for the period $(t-2)$ is given by

$$f_{t-2} = \theta \, d_{t-3} + (1 - \theta) \, f_{t-3} \tag{10.73}$$

where

d_{t-3} is the actual demand for the time period $(t-3)$.
f_{t-3} is the forecast for the time period $(t-3)$.

Substituting Equation (10.73) into Equation (10.72) results in

$$f_{t-1} = \theta \, d_{t-2} + [1 - \theta][\theta d_{t-3} + (1 - \theta)f_{t-3}]$$

$$= \theta \, d_{t-2} + \theta(1 - \theta) \, d_{t-3} + (1 - \theta)^2 \, f_{t-3} \tag{10.74}$$

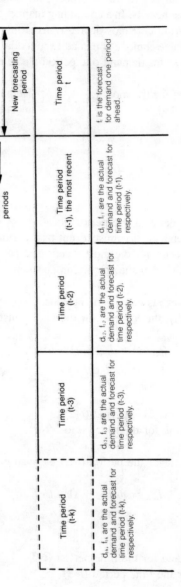

Figure 10.4.
Equal time periods.

Similarly, substituting Equation (10.74) into Equation (10.71) yields

$$f_t = \theta \, d_{t-1} + [1 - \theta][\theta d_{t-2} + \theta(1 - \theta)d_{t-3} + (1 - \theta)^2 f_{t-3}]$$

$$(10.75)$$

$$f_t = \theta \, d_{t-1} + \theta(1 - \theta) \, d_{t-2} + \theta(1 - \theta)^2 \, d_{t-3} + (1 - \theta)^3 f_{t-3}$$

The above equation leads to the following generalized equation:

$$f_t = \theta(1 + \theta)^0 d_{t-1} + \theta(1 - \theta)^1 d_{t-2} + \theta(1 - \theta)^2 d_{t-3} + ---$$

$$(10.76)$$

$$+ \; \theta(1 - \theta)^{k-1} d_{t-k} + (1 - \theta)^k f_{t-k}$$

where k is the number of past periods.

It can be seen from Equation (10.76) that the weights associated with each observation of Equation (10.76) are not equal but rather the successively older observations weights decrease by factor $(1 - \theta)$. In other words, the successive terms $\theta(1 - \theta)^0$, $\theta(1 - \theta)^1$, $\theta(1 - \theta)^2$, $\theta(1 - \theta)^3$, $---$ decrease exponentially. This means the more recent observations are more heavily weighted than the remote observations.

The frequently used values for θ are between 0.01 and 0.3. At $\theta = 1$ in Equation (10.76), the consumption or demand of the last period is the forecast for the period ahead.

Similarly, for the very low value of θ (in other words, the value of θ close to zero), Equation (10.76) gives almost equal weights to all previous observations.

EXAMPLE 10.17

For the months of March, April, May, June and July an electronics manufacturer sold 40, 30, 50, 40 and 60 units, respectively, of a specific product. If the value of $\theta = 0.2$ and the forecast for March was 50 units, forecast the number of units to be sold in August by utilizing the exponential smoothing technique.

By substituting the specified data into Equation (10.71), the forecast for April (i.e., $t = $ April) is

$$f_t = 0.2(40) + (1 - 0.2)(50) = 48 \text{ units}$$

where

(40) = Actual consumption or demand for March.
(50) = Forecast for March.

Similarly, substituting the specified data into Equation (10.71) leads to the resulting forecast for May (i.e., $t = $ May):

$$f_t = 0.2(30) + (1 - 0.2)(48) = 44.4 \text{ units}$$

where

(30) = Actual consumption or demand for April.
(48) = Forecast for April.

In similar fashion the forecasts for the remaining months were obtained as given in Table 10.1.

Thus the forecast for the month of August is 47.536 units.

The same result can be obtained by substituting the specified data into Equation (10.76). For example, the forecast for the month of August (i.e., $t = $ August) is

$$f_t = \theta(1 + \theta)^0 d_{t-1} + \theta(1 - \theta)^1 d_{t-2} + \theta(1 - \theta)^2 d_{t-3} + \theta(1 - \theta)^3 d_{t-4}$$

$$+ \theta(1 - \theta)^4 d_{t-5} + (1 - \theta)^5 f_{t-5}$$

$$= (0.2)(60) + (0.16)(40) + (0.128)(50) + (0.1024)(30)$$

$$+ (0.0819)(40) + 0.328 (50)$$

$$= 47. 548 \text{ units.}$$

10.7 DECISION TREES

The decision tree analysis is used to deal with sequential problems. A decision tree may be simply described as a schematic diagram of a sequence of alternative decisions as well as the conclusions of those decisions [11].

Table 10.1. Forecasts for Various Months.

Month	Actual demand (units)	Forecast (units)
May	50	44.4
June	40	45.52
July	60	44.42
August	—	47.536

There are various advantages of the decision tree. Some of them are as follows:

(i) It can take into consideration the actions of more than one decision maker.

(ii) It simplifies the expected value calculations.

(iii) It is a visibility tool to represent the sequential decision process.

To perform the decision tree analysis there are basically three steps involved [10]. These are tree diagramming, estimation, and evaluation and selection. Tree diagramming is composed of three steps. These are:

(i) Identification and sequence of the decisions and their alternatives

(ii) Identification of chance events

(iii) Construction of the tree diagram depicting decisions and chance events sequence

The estimation aspect is concerned with estimating the chance event occurrence probability and financial consequences of all possible outcomes.

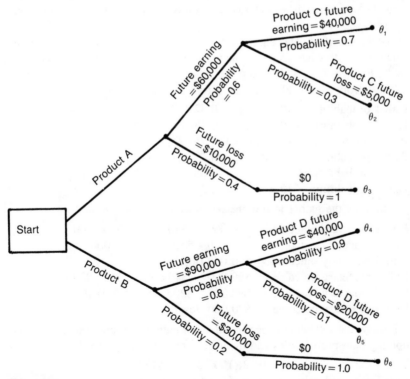

Figure 10.5.
A decision tree.

Finally, the evaluation and selection are concerned with computing the actions' expected values and choosing the action with the best expected value. The concept of the decision tree is described in the following example [12].

EXAMPLE 10.18

A small engineering company has an option to develop either type A or type B product. After a careful analysis, it is felt that there is about a 60% chance that product A will bring a total revenue of $60,000 and a 40% chance that by selecting product A for development, the company will lose $10,000. Furthermore, if product A is a success then the company can also develop product C which has a 70% chance of earning $40,000 and a 30% chance of losing $5,000.

However, in the case of product B there is an 80% chance of earning $90,000 and a 20% chance of losing $30,000. In the event product B is successful, the company has an opportunity to develop another product, D, which has a 90% chance of earning $40,000 and a 10% chance of losing $20,000. Product C and D will be produced only if products A and B, respectively, are successful. Otherwise, these products (C and D) will not be developed. Therefore, the company wishes to know which product it should consider for development.

For the specified data, Figure 10.5 shows a newly constructed decision tree. The decision tree has six paths and the ith path determination point is denoted by θ_i, for $i = 1, 2, 3, 4, 5, 6$.

Thus, from Figure 10.5 the path 1 to 6 probabilities are:

Path 1 probability $= (0.6)(0.7) = 0.42$
Path 2 probability $= (0.6)(0.3) = 0.18$
Path 3 probability $= (0.4)(1) = 0.4$
Path 4 probability $= (0.8)(0.9) = 0.72$
Path 5 probability $= (0.8)(0.1) = 0.08$
Path 6 probability $= (0.2)(1.0) = 0.2$

Similarly, the path 1 to 6 monetary outcomes are as follows:

Path 1 monetary outcome $= \$60,000 + \$40,000 = \$100,000$
Path 2 monetary outcome $= \$60,000 - \$5,000 = \$55,000$
Path 3 monetary outcome $= \$10,000$ (loss)
Path 4 monetary outcome $= \$90,000 + \$40,000 = \$130,000$
Path 5 monetary outcome $= \$90,000 - \$20,000 = \$70,000$
Path 6 monetary outcome $= \$30,000$ (loss)

Finally, the following expected values of path 1 to 6 are obtained by multiplying path probabilities and monetary outcomes, respectively:

Path 1 expected value $= (0.42)(100,000) = \$42,000$
Path 2 expected value $= (0.18)(55,000) = \$9,900$
Path 3 expected value $= (0.4)(10,000) = \$4,000$

Path 4 expected value = (0.72)(130,000) = \$93,600
Path 5 expected value = (0.08)(70,000) = \$5,600
Path 6 expected value = (0.2)(30,000) = \$6,000

By utilizing the above result, the expected value, *EV*, of the decision to develop product A is

$$EV = 42,000 + 9,900 - 4,000 = \$47,900$$

Similarly, the expected value, *EV*, of the decision to develop Product B is

$$EV = 93,600 + 5,600 - 6,000 = \$93,200$$

Thus, it would be wise to develop product B because the company will be better off by \$45,300 (i.e., \$93,200 − \$47,900).

10.8   FAULT TREES

This is a technique widely used in industry to evaluate reliability of complex engineering systems. Therefore, it is felt that an engineering manager must have some basic knowledge of this important technique, even though the engineering manager may not be directly concerned with the reliability evaluation of technical systems. Furthermore, this technique is briefly described here because of its schematic resemblance and name similarity to the decision tree concept. In addition, probability theory is utilized in both techniques.

In the early sixties, the fault tree technique was originated in Bell Laboratories to evaluate reliability of the Minuteman Launch Control System.

The fault tree analysis starts by identifying an undesirable event, called the top event, associated with a system. Events which could cause the occurrence of a top event are generated and connected by logic gates known as the AND, OR and so on. The construction of a fault tree proceeds by generation of fault events in a successive manner until the fault events need not be developed further. These fault events are known as primary events. In simple terms, the fault tree may be described as the logic structure relating the top event to the primary events.

10.8.1   Fault Tree Symbology

This section presents the basic symbols associated with the fault tree concept. More comprehensive symbols may be found in Reference [13].

CIRCLE

This is used to represent the failure of an elementary component (basic fault event). The elementary component failure parameters such as constant failure

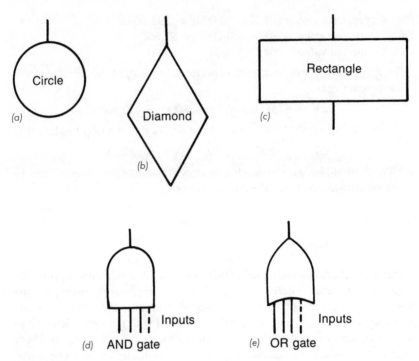

Figure 10.6.
Fault tree symbols.

and repair rates, probability and unavailability are calculated from the field data or other sources. The symbol is shown in Figure 10.6(a).

DIAMOND

This symbol is used to represent a fault event whose causes will not be fully developed due to lack of interest or data. The diamond symbol is shown in Figure 10.6(b).

RECTANGLE

This symbol is used to represent an event which occurs due to combination of fault events through the input of a gate. The rectangle symbol is given in Figure 10.6(c).

AND GATE

This gate is used to represent a situation such that an output fault event will occur only if all the input fault events occur. The AND gate symbol is shown in Figure 10.6(d).

OR GATE

This is the opposite of the AND gate. Therefore, the OR gate signifies that an output fault event occurs if one or more of the input fault events occur. The OR gate symbol is shown in Figure 10.6(e).

10.8.2 Fault Tree Construction

Before a reliability analyst starts constructing a system fault tree, he has to thoroughly understand the system under study. Once this is accomplished, the analyst starts the fault tree construction to represent system conditions (utilizing the specified symbols) which may cause the system to fail. There are various advantages of the fault tree construction. Some of them are as follows:

(i) Identifies system weakness in a visible form

(ii) Acts as a visual tool in communication

(iii) Evaluates the system design adequacy

(iv) Acts as a visual tool to perform trade of studies

Fault tree construction is demonstrated with the aid of the following example.

EXAMPLE 10.19

An engineering workshop room has four light bulbs controlled from a single switch. The room will be dark only when all four light bulbs fail, the switch fails to close, or the power is off. Construct the fault tree for this simple system by assuming that the dark room is the top event.

Thus, Figure 10.7 shows a fault tree for the top event being a dark room. The fault tree basic fault events (A, B, C, D, E, F and G) are denoted by circles, and intermediate and top fault events by rectangles. The room can be dark only if either of the following happens:

(i) There is no power.

(ii) All bulbs fail.

(iii) Light switch fails to close.

Thus to represent either situation, the above three conditions provide input to an OR gate, G_1. Furthermore, investigating the no-power possibility, event X, further indicates that it can happen only if power or fuse fails. Therefore these two events (A,B) provide input to an OR gate, G_2. Finally, further investigation of bulb failure, event Y, shows that it can occur only if all the four light bulbs fail. Thus the bulb failure events C, D, E and F provide input to an AND gate, G_3. The room will be dark if fault event A or B or Y or G occurs.

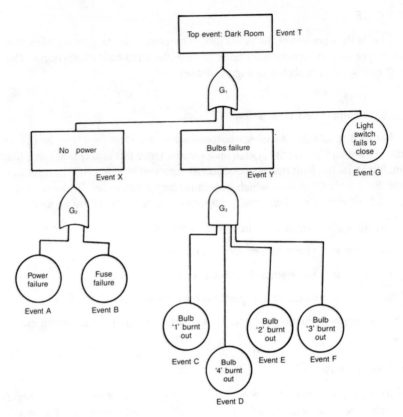

Figure 10.7.
Fault tree for dark room.

10.8.3   Quantitative Analysis of Fault Trees

This section is concerned with the evaluation of probability of occurrence of top fault event. Basic probability theory is used to compute the probabilities of OR and AND gates top event occurrence.

OR GATE

A "k" independent input fault event OR gate is shown in Figure 10.8. The occurrence probability of event ET is given by

$$P(ET) = 1 - \prod_{j=1}^{k} (1 - p_j) \qquad (10.77)$$

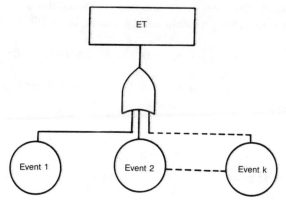

Figure 10.8.
A "k" inputs OR gate.

where

$P(ET)$ is the occurrence probability of event ET.
k is the number of OR gate input events.
p_j is the occurrence probability of jth input fault event; for $j = 1,2,3,$
       $--- k$.

However, if the input event probabilities are very small, then Equation (10.77) reduces to

$$P(ET) \cong \sum_{j=1}^{k} p_j \qquad (10.78)$$

EXAMPLE 10.20

An OR gate has four independent inputs. The input fault events 1, 2, 3 and 4 probabilities of occurrence are 0.01, 0.02, 0.01 and .03, respectively. Calculate the OR gate top (output) event probability of occurrence.

For four inputs (i.e., $k = 4$), Equation (10.77) reduces to

$$P(ET) = 1 - (1 - p_1)(1 - p_2)(1 - p_3)(1 - p_4) \qquad (10.79)$$

Substituting the specified data for p_1, p_2, p_3 and p_4 into Equation (10.80) leads to

$$P(ET) = 1 - (1 - 0.01)(1 - 0.02)(1 - 0.01)(1 - 0.03)$$

$$= 0.068317$$

In this example the input event occurrence probabilities are small. Thus substituting the specified probabilities data into Equation (10.78) results in

$$P(ET) \cong p_1 + p_2 + p_3 + p_4 = 0.01 + 0.02 + 0.01 + 0.03 = 0.07$$

In this case, it is demonstrated that Equations (10.77) and (10.78) yield almost the same end result. The probability of OR gate top event occurrence is approximately 0.07.

AND GATE

An m independent input fault event AND gate is shown in Figure 10.9. Probability of occurrence of top event, ET, is given by

$$P(ET) = \prod_{j=1}^{m} p_j \qquad (10.80)$$

where

$P(ET)$ is the probability of occurrence of top event ET.
m is the number of AND gate input events.
p_j is the occurrence probability of jth input fault event; for $j = 1, 2, 3, ---, k$.

EXAMPLE 10.21

An AND gate has two independent inputs. Each input fault event occurrence probability is 0.05. Compute the AND gate top (output) event occurrence probability.

Thus substituting the specified data into Equation (10.80) yields

$$P(ET) = p_1 \cdot p_2 = (0.05)(0.05) = 0.0025$$

The AND gate top (output) event probability of occurrence is 0.0025.

FAULT TREE EVALUATION

Here we are concerned with calculating the probability of occurrence of a fault tree top event. The AND and OR gate probability evaluation Equations (10.77), (10.78), and (10.80) are used to evaluate the occurrence probability of the fault tree top event. However, before applying Equations (10.77), (10.78) and (10.80), the reliability analyst has to make certain that there is no repeated event in the fault tree diagram. The fault tree top event probability of occurrence evaluation is demonstrated in the following example.

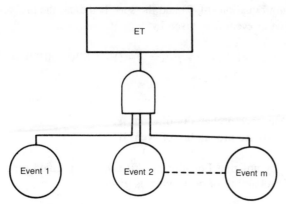

Figure 10.9.
An "m" inputs AND gate.

EXAMPLE 10.22

In Figure 10.7 the probability of occurrence of independent events A, B, C, D, E, F and G are 0.01, 0.02, 0.03, 0.03, 0.03, 0.03 and 0.02, respectively. Calculate the probability of having a dark room (i.e., top event, T).

In Figure 10.7 there is no repeated event; therefore Equations (10.77), (10.78) and (10.80) can be used to evaluate top event occurrence probability. The Figure 10.7 fault tree with probability calculations is redrawn in Figure 10.10.

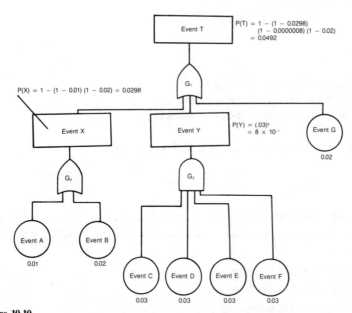

Figure 10.10.
A fault tree (for dark room) with probability calculations.

By utilizing Equation (10.77) and the specified data, the probability, $P(X)$, of occurrence of event X is given by

$$P(X) = 1 - (1 - p_1)(1 - p_2) = 1 - (1 - 0.01)(1 - 0.02) = 0.0298$$

where

p_1 is the occurrence probability of event A.
p_2 is the occurrence probability of event B.

Similarly, by utilizing Equation (10.80) and the specified probability data, the event Y probability of occurrence is given by

$$P(Y) = p_1 \cdot p_2 \cdot p_3 \cdot p_4 = (0.03)(0.03)(0.03)(0.03) = 0.0000008$$

where p_1, p_2, p_3, p_4 are the occurrence probabilities of events C, D, E and F, respectively.

Finally, utilizing Equation (10.77) computed results and specified data, the top event T occurrence probability is

$$P(T) = 1 - (1 - p_1)(1 - p_2)(1 - p_3)$$

$$= 1 - (1 - 0.0298)(1 - .0000008)(1 - 0.02)$$

$$= 0.0492$$

where

p_1 is the occurrence probability of event X.
p_2 is the occurrence probability of event Y.
p_3 is the occurrence probability of event G.

Thus the probability of having a dark room is 0.0492.

10.9 SUMMARY

This chapter briefly describes various techniques used to make better engineering management decisions. The following items are covered in the chapter:

(i) Optimization techniques
(ii) Discounted cash flow analysis
(iii) Depreciation techniques

(iv) Business operations analysis

(v) Forecasting

(vi) Decision trees

(vii) Fault trees

Under the section on optimization techniques, the two commonly known techniques are described with the aid of three numerical examples. These techniques are Lagrangian multiplier and linear programming. In the discounted cash flow analysis, simple interest, compound interest, present worth, and uniform periodic payment models are discussed by solving several numerical examples.

Depreciation techniques such as declining-balance, straight-line depreciation and sum-of-digits are presented with the aid of various solved numerical examples. Another topic of the chapter is the business operation analysis. Under this heading, one mathematical model is presented. The model is concerned with finding the point of maximum profit, the point of maximum investment rate, and the point of maximum economic production.

Forecasting is another area which is covered in this chapter. Empirical and analytical forecasting approaches are described. Another technique frequently used to make decisions concerning sequential problems is the method of decision trees. This method is briefly discussed.

Finally, the technique known as the fault tree is presented. This technique is concerned with reliability evaluation of engineering systems.

The chapter contains 22 solved examples. The source references for the material described in the chapter are listed at the end of the chapter.

10.10 EXERCISES

1. Determine the critical point for $f(x_1,x_2) = x_1^2 + x_2^2$ subject to
   $$k(x_1,x_2) = x_1 - x_2 - 6 = 0$$

2. Maximize $z = 20x + 30y$
   subject to
   $$15x + 5y \leq 80$$
   $$5x + 6y \leq 30$$
   $$x \geq 0$$
   $$y \geq 0$$

3. A mechanical component manufacturer has borrowed $1,000,000 at a compound interest rate of 10% per annum. The borrowed sum of money has to be fully paid back at the end of the five-year period. Determine the total compound interest to be paid at the end of the specified period.

4. A company has recently purchased a machine. The expected useful life of the machine is 10 years. After the useful life period, the machine salvage value will be $50,000. If the compound interest rate is 10% per annum, calculate the present worth of the salvage value.

5. A manufacturer has recently procured a piece of equipment for $90,000. After a seven-year useful life period in service, the equipment is expected to be sold for $40,000. At the end of each year the equipment estimated maintenance cost is $2,000. If the compound interest rate is 12% per annum, compute the present value of the equipment net total cost.

6. An engineering company has invested $150,000 to purchase a mechanical machine. At the end of a 10-year useful life period, the estimated salvage value of the machine is $20,000. If the machine annual depreciation is constant, calculate the machine annual depreciation charge.

7. Describe the difference between the empirical and analytical forecasting techniques with respect to their application areas.

8. Compare the simple moving average model with the exponential smoothing model.

9. The monthly sales data for a specified product were as tabulated in Table 10.2. For the month of February the sales of 135 units were forecasted. If the value of the smoothing constant is 0.3, forecast the product sales for October.

Table 10.2. Monthly Sales Data for a Certain Product.

Month	Units sold
February	140
March	130
April	160
May	170
June	140
July	150
August	170
September	130

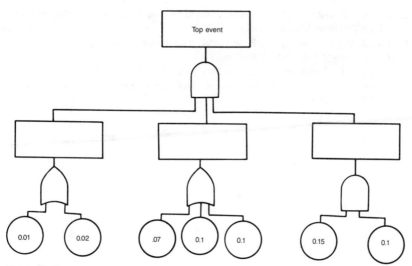

Figure 10.11.
A hypothetical fault tree.

10. Discuss the similarities and differences of decision trees and fault trees.

11. Determine the occurrence probability of top event of the fault tree given in Figure 10.11. The independent basic fault event probabilities are specified in the fault tree diagram.

10.11   REFERENCES

1. Gue, R. L. *Mathematical Methods in Operations Research*. London:Collier-Macmillan Limited (1968).

2. Dantzig, G. B. *Linear Programming and Extensions*. Princeton, NJ:Princeton University Press (1963).

3. Howell, J. E. and D. Teichroew. *Mathematical Analysis for Business Decisions*. Homewood, IL 60430:Richard D. Irwin, Inc. (1971).

4. DeGarmo, E. P., J. R. Canada and W. G. Sullivan. *Engineering Economy*. New York:Macmillan Publishing Co., Inc. (1979).

5. Dhillon, B. S. *Reliability Engineering in Systems Design and Operations*. New York:Van Nostrand Reinhold Company (1982).

6. Riggs, J. L. *Economic Decision Models for Engineers and Managers*. New York:McGraw-Hill Book Company (1968).

7. Kurtz, M. *Engineering Economics for Professional Engineers' Examinations*. New York:McGraw-Hill Book Company (1975).

8. Laufer, A. C. *Operations Management*. Cincinnati:South-Western Publishing Co. (1975).

9. Riggs, J. L. *Production Systems: Planning, Analysis and Control*. New York:John Wiley & Sons (1981).

10. Adam, E. E. and R. J. Ebert. *Production and Operations Management*. Englewood Cliffs, NJ 07632:Prentice-Hall, Inc. (1982).

11. Lee, S. M., L. J. Moore and B. W. Taylor. *Management Science*. Dubuque, IA:Wm.C. Brown Company Publishers (1981).

12. Tellier, R. D. *Operations Management*. New York:Harper & Row (1978).

13. Schroder, R. J. "Fault Tree for Reliability Analysis," *Proceedings of the Annual Symposium on Reliability*. Published by IEEE (1970).

CHAPTER 11

Mathematical Models for Engineering Management Decision Making

11.1 INTRODUCTION

In today's competitive environments, various types of mathematical models are used to provide useful input factors in engineering management decision making. These decisions may be concerned with equipment or product production, procurement, repair, etc. When utilizing a mathematical model, whether for management or engineering decision making, care must be given particularly to the assumptions associated with the model and the accuracy of the input data to that model. Otherwise the inputs from these models will lead to incorrect conclusions.

The main objective of this chapter is to familiarize the reader with selective mathematical models which are not generally available in the common literature on engineering management. The models in this chapter basically fall into the following three areas:

(i) Cost-capacity

(ii) Financial investment modeling

(iii) Equipment repair

These models are presented in the following sections.

11.2 LARGE PLANT INVESTMENT DECISION MODELS

This section presents two mathematical models as follows.

11.2.1 Model I

This model is known as the cost-capacity model [1–3]. In this model it is assumed that for equipment or a plant under consideration, cost data on similar equipment or plant of different capacity is available. Thus the cost of new equipment or plant, k_n, under consideration may be calculated by utilizing the following equation:

$$k_n = k_s \cdot (C_n)^{\beta} \cdot (C_s)^{-\beta}$$

(11.1)

where

k_s is the known cost of the similar plant or equipment of capacity C_s.
C_s is the known capacity of the similar plant or equipment.
C_n is the capacity of the new plant or equipment under consideration.
β is the cost-capacity factor [2–3].

According to Reference [1], the value of the cost-capacity factor varies from less than 0.2 to more than unity. However, on the average, the value of β factor is 0.6. In the event that no information is available on the value of the capacity factor then the average value of 0.6 should be used. The values of the cost-capacity factor for various types of equipment are tabulated in Reference [1].

EXAMPLE 11.1

The estimated cost of a 500 MW electric generator is $5 million. A company is to build a 1,000 MW generator. Calculate the cost of the new generator by assuming that the value of the cost-capacity factor is 0.8.
Thus utilizing the given data in Equation (11.1) we get:

$$k_n = 5(1,000)^{0.8} (500)^{-0.8}$$

$$= \$8.706 \text{ million}$$

11.2.2   Model II

This model is known as the cost index which is used for comparison purposes, for example, when one is comparing from one year to another the cost or price changes for a specified quantity of products or services. The history of the index numbers goes back to the middle of the 18th century when they were used by an Italian, G. R. Carli [4]. A cost index for a specified year or time shows the cost at that time in relation to a given base year. Therefore it is simply a dimensionless number. Thus, the following formula is used to predict present, future, or past cost:

$$k_{pfp} = k_r \cdot I_{pfp}/I_r \tag{11.2}$$

where

k_r denotes the original reference cost.
k_{pfp} denotes the present, future or past cost.
I_r denotes the value of the index. This index represents that value when the value of the reference cost was obtained.
I_{pfp} denotes the present, future or past time value of the index.

The main advantage of this cost index is that it allows the estimator to predict the cost of the similar product design from the past time to the present or future period, without going into extensive cost analysis.

To estimate the value of the index, I_{pfp}, the commonly used method is known as the weighted arithmetic technique. Thus for m items, the index, I_{pfp}, is defined as follows:

$$I_{pfp} = \frac{1}{m} \sum_{i=1}^{m} (k_{1i}/k_{0i}) \tag{11.3}$$

where

k_{1i} denotes, for the first year, the cost of the ith item; for $i = 1,2,3,---,m$.
k_{0i} denotes, for the base year (zero year), the cost of the ith item; for $i = 1,2,3,---,m$.
m denotes the number of items.

The Equation (11.3) index is known as the simple index because it treats all items in question equally. However, in real life, some items may be more important than others. Therefore, to take this possibility into consideration, Equation (11.3) is modified by assigning a weight to each item as follows:

$$I_{pfp} = \left[\sum_{i=1}^{m} (w_i k_{1i})/k_{0i} \right] \Bigg/ \sum_{i=1}^{m} w_i \tag{11.4}$$

where w_i denotes the ith item weight; for $i = 1,2,3,---,m$.

Other indexes are covered in Reference [6].

EXAMPLE 11.2

An electric utility is to build a 5000 megawatts electric power generating station. A few years ago a similar power plant was constructed. The per megawatt construction cost was $250,000 and the value of the index was 125. The forecasted value of the index for the proposed construction period is 150. Calculate the per megawatt proposed power plant construction cost.

In this example the data for the following item are specified:

$$k_r = \$250,000, \; I_r = 125, \; I_{pfp} = 150$$

Thus substituting the above specified data into Equation (11.2) leads to

$$k_{pfp} = (250,000)(150)/(125) = \$300,000$$

The power plant construction cost per megawatt will be $300,000.

11.3 FINANCIAL INVESTMENT MODELING

This section presents one mathematical model to determine the optimum area to be served by a warehouse. The model is taken from Reference [6]. It is generally accepted that as the volume of goods in a warehouse increases, the warehousing costs per dollar's worth of goods decreases. Therefore, it is fair to conclude that a large warehouse is more cost efficient. In this model, it is assumed that a company has a number of warehouses. If it increases the size of its warehouses, the number of warehouses required and the goods warehousing costs would be reduced in turn. However, it will result in increasing the goods delivery and transportation costs, because the company delivery trucks have to cover greater distances to deliver stored goods to the buyers. Thus, the cost, k, per dollar's worth of goods distributed in the warehouse district boundaries is given by

$$k = \sum_{i=1}^{3} k_i \qquad (11.5)$$

where

k_1 denotes the operating costs of the warehouse. (This is the cost per dollar's worth of goods delivered.)
k_2 denotes the goods delivery costs.
k_3 denotes the fixed costs (i.e., these are independent of volume of goods and area served and are given as per dollar's worth of goods).

However, k_1 and k_2 are given by

$$k_1 = \frac{k_f}{v} \qquad (11.6)$$

where

k_f denotes the warehouse operation fixed costs.
v denotes volume of goods per unit of time (i.e., month, year, etc.).

The volume must be given in dollars and

$$k_2 = k_v(a)^{1/2} \qquad (11.7)$$

where

a denotes the area of the district, in square miles, to which warehouse distributes its goods.

k_v denotes that cost which is assumed to vary with $\sqrt{a}$. (This cost occurs due to truck breakdowns and repairs, gasoline, driver salary, etc.)

The relationship, R (constant dollar volume of goods sold per square mile), between volume and the area served is defined by

$$R = v/a \qquad (11.8)$$

Thus substituting from Equation (11.8) for v into Equation (11.6) leads to

$$k_1 = k_f/a\,R \qquad (11.9)$$

Incorporating Equations (11.7) and (11.9) into Equation (11.5) yields

$$k = \frac{k_f}{aR} + k_v\,a^{1/2} + k_3 \qquad (11.10)$$

In order to minimize the cost, k, differentiate Equation (11.10) with respect to a and set the resulting expression equal to zero as follows:

$$\frac{dk}{da} = -\frac{k_f}{a^2 R} + \frac{1}{2}\,k_v\,a^{-1/2} = 0 \qquad (11.11)$$

Rearranging Equation (11.11) leads to the following equation for the optimum value

$$a^* = (2\,k_f/R\,k_v)^{2/3} \qquad (11.12)$$

11.4 ENGINEERING EQUIPMENT REPAIR FACILITY DECISION MODELS

This section presents two queuing problem models which can be used to make engineering equipment repair facility decisions. These models are as follows.

MODEL I

This is known as the single channel, single phase queuing model with infinite population. The repair time, t_1, to repair a single piece of equipment is less than the time between arrivals, t_2. Because of no queue forms, the lost time or the waiting time has the same value as the servicing time, t_1. Thus, the total variable cost, k_t, per unit time from Reference [1] is given by

$$k_t = \sum_{i=1}^{2} k_i = \frac{k_s}{t_1} + \frac{t_1}{t_2} \cdot k_{wt} \qquad (11.13)$$

where

k_1 denotes the repair (servicing) cost per unit time.

k_2 denotes the waiting cost per unit time.

k_s denotes the repair (servicing) cost of a single piece of equipment. This cost does not include the cost of materials used for repair but includes the cost for maintaining the repair facility.

k_{wt} denotes the cost of waiting per piece of equipment per unit time.

By assuming that the repair time is an independent variable, we differentiate Equation (11.13) with respect to t_1 as follows:

$$\frac{dk_t}{dt_1} = -\frac{k_s}{t_1^2} + \frac{k_{wt}}{t_2} \tag{11.14}$$

To obtain the minimum value of k_t by setting the derivative in Equation (11.14) equal to zero yields:

$$\frac{k_{wt}}{t_2} - \frac{k_s}{t_1^2} = 0 \tag{11.15}$$

Thus, rearranging Equation (11.15) results in:

$$t_1^* = \left(\frac{k_s t_2}{k_{wt}}\right)^{1/2} \tag{11.16}$$

where t_1^* is the optimum repair or service time.

Substituting Equation (11.16) into Equation (11.13) leads to the optimum value of

$$k_t = 2\left(\frac{k_s \cdot k_{wt}}{t_2}\right)^{1/2} \tag{11.17}$$

EXAMPLE 11.3

An equipment repair facility can be represented by a single channel, single phase queuing model with infinite population. The time between failed equipment arrivals is equal to 0.5 hours. When a failed piece of equipment waits for repair, it costs \$150/hour. Furthermore, it costs \$50 to repair a single piece of failed equipment. (This also includes the cost of maintaining the repair facility.) Calculate the value of the optimum time to repair a single piece of equipment.

In this example, the data is given for the following items:

$$t_2 = 0.5 \text{ hours}, \ k_s = \$50/\text{equipment}, \ k_{wt} = \$150/\text{hour}$$

Utilizing the above given data in Equation (11.16) yields:

$$t_1^* = \frac{(50)(0.5)}{150} = 0.1667 \text{ hour}$$

Thus, the value of the optimum time to repair a single piece of equipment is equal to 0.1667 hour.

MODEL II

This model is the result of further development on Model I. However, the only difference between this model and Model I is that this one has M service or repair channels instead of only one. Thus, one can write directly the equation for the M channel arrangement total variable cost, k_{mt}, per unit time from Equation (11.13) as follows:

$$k_{mt} = \frac{k_s M}{t_1} + k_{wt} \cdot \theta \tag{11.18}$$

where

$$\theta \equiv t_1 / t_2 \tag{11.19}$$

Differentiating Equation (11.18) with respect to t_1 yields:

$$\frac{d \, k_{mt}}{d \, t_1} = \frac{k_{wt}}{t_2} - \frac{k_s M}{t_1^2} \tag{11.20}$$

Setting the derivative of Equation (11.20) equal to zero results in:

$$\frac{k_{wt}}{t_2} - \frac{k_s M}{t_1^2} = 0 \tag{11.21}$$

Rearranging Equation (11.21) leads to

$$t_1^* = \sqrt{\frac{k_s M t_2}{k_{wt}}} \ , \text{ for } M t_2 \geq t_1 \tag{11.22}$$

where t_1^* denotes the optimum value of t_1.

Substituting Equation (11.22) for t_1 into Equation (11.18) yields

$$k_{mt}^* = 2 \left(\frac{k_s k_{wt} M}{t_2} \right)^{1/2} \qquad (11.23)$$

where k_{mt}^* denotes the optimum value of k_{mt}.

11.5 SUMMARY

This chapter presents five selective mathematical models which will be useful in making better engineering management decisions. These models are concerned with making large investment decisions and equipment repair shop decisions. The first two models are basically based on the cost-capacity concept. These models make use of the data on cost and capacity of the old but similar project to predict the cost of the new project. The third model is concerned with finding the optimum area to be served by a warehouse. The last two models deal with finding the optimum time to repair a single piece of equipment. The basis for these models are the queuing arrangements of single channel, single phase and multiple channels, single phase with inifinite population, respectively.

The chapter contains three solved examples. The source references for the material presented are listed at the end of the chapter.

11.6 EXERCISES

1. A steel plant of size 500,000 tons/year costs $50 million to build. Calculate the cost of a 2,000,000 tons/year plant.

2. A steel company is to build a steel plant of capacity of 5 million tons/year. A similar plant was built in the past when the construction cost was $50,000/ton and the index value was 120. The estimated value of the index for the proposed plant construction period is 140. Estimate the cost of the proposed plant on the per ton basis.

3. A single channel, single phase model with infinite population describes an equipment repair facility. To repair a failed piece of equipment costs $300. This cost also includes the repair facility maintenance cost. It cost $400/hour when a failed piece of equipment has to wait for repair. In addition, the time between failed equipment arrivals is 40 minutes. Find the optimum time to repair a single piece of equipment.

11.7 REFERENCES

1. Jelen, F. C. ed. *Cost and Optimization Engineering.* New York:McGraw-Hill Book Company (1970).
2. Chilton, C. H. "Six-Tenths Factor Applies to Complete Plant Costs," *Chemical Engineering,* 57:112–114 (April 1950).
3. Williams, R. "Six-Tenths Factor Aids in Approximating Costs," *Chemical Engineering,* 54:124–125 (December 1947).
4. Ostwald, P. F. *Cost Estimating for Engineering and Management.* Englewood Cliffs, NJ:Prentice-Hall, Inc. (1974).
5. For a detailed summary of cost indexes, see *Engineering News Record,* 178(11):87–163 (1967).
6. Bowman, E. H. and J. B. Stewart, "A Model for Scale of Operations," *Journal of Marketing,* 20:242–247 (January 1956).

CHAPTER 12

Engineering Product Developing and Costing

12.1 INTRODUCTION

This chapter briefly describes two important topics which are of day-to-day concern to any engineering manager.

Evidence from the past indicates that growth of an engineering equipment manufacturer largely depends on the development of new products. Despite these indications, in many organizations the planning for the development of new products seems to be a weak spot. This may be the reason for an extremely high percentage of new product failure in the market. For example, according to Reference [1], in the United States more than 98% of the products cannot survive the first 24 months. Usually there are two procedures practiced to establish corporate objectives in regard to product planning. In the first approach, the product planning objectives are established at the vice-presidential level or higher. However, in the second procedure the committee concept is followed, especially in bigger companies.

Product costing is another aspect which is very important to management, because any engineering company would venture into the development of a new product only if it were going to generate profit to that company. In order for a company to estimate profit, the cost of various aspects associated with the product under consideration have to be estimated. However, there are various other reasons for product costing which are described in the chapter.

Both the topics of engineering product developing and costing are described separately in the chapter.

12.2 PRODUCT DEVELOPING

As mentioned earlier, the development of new products is largely responsible for the growth of the engineering organizations. There are various risks associated with the development of new products.

According to References [2,3], in 1971 the estimated expenditure to develop new products in the United States alone was approximately $22.4 billion. According to Spitz [3], most of this money was spent on those new product ideas which would ultimately be unsuccessful. This belief is further justified by the study results reported by Buzzell and Nourse [4,3]. According to their find-ings, in the food industry alone out of every 1000 new product ideas only 43 of them are developed and put into the market. However, after the introduction of the product to the market only 36 remain on the market. Furthermore, both researchers point out that 810 out of 1000 new food product ideas are rejected at the idea stage and 135 out of the remaining ideas are discontinued at the product testing stage. Another 12 of them are rejected after conducting the marketing tests.

These studies certainly lead one to believe that there is a definite need for extreme carefulness when developing new engineering products. Therefore this section briefly explores various aspects associated with developing new engineering products.

12.2.1   Important Reasons for Developing New Engineering Products

There are various reasons to develop new engineering products. However, generally the ultimate objective of all these reasons is to increase profit. Some of these reasons are:

(i) *Excess Capacity:* This is concerned with the company having an excess capacity in areas such as production, marketing, etc. To utilize such areas effectively, the company develops new products. The main idea behind this strategy is that the fixed costs will spread over a larger quantity of units.

(ii) *Utilization of By-Products:* In this case the new products are developed to make use of waste or by-products left over from other products. The petroleum industry is one example which makes use of by-products to develop other products.

(iii) *Competition:* This is another incentive for introducing new products. Many companies introduce new products just to hold their market. Otherwise, it may not be possible in the present competitive environments.

(iv) *Seasonal Fluctuations:* Companies whose manufactured goods sales are subject to seasonal variations smooth out these changes by introducing new products.

(v) *Phased-Out Products:* In this situation the new products are developed to replace the phased-out products, because many companies terminate the production of products which bring in a very low profit.

(vi) *New Opportunities:* Many new products are developed to exploit new opportunities.

(vii) *Risk:* This is another important reason to produce new products. Here, the objective of developing new products is to reduce risk faced by the company. An example of such a risk is when a small percentage of a firm's products account for a very large percentage of overall total sales volume of that company.

12.2.2 Steps Involved in New Engineering Product Development

According to Reference [6] there are essentially six steps involved in developing new engineering products. These six steps are shown in Figure 12.1

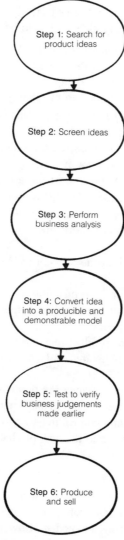

Figure 12.1.
Steps for developing new engineering products.

which is concerned with idea exploration, idea screening, business analysis, development of a model, product testing and product commercialization.

It is to be noted that in Figure 12.1 the cost of developing an engineering product increases as progress is made from one step to another (i.e., say from step 1 to step 2, from step 2 to step 3, etc).

12.2.3 Tasks Useful in Effectively Managing the New Engineering Product Development

Before the company management can effectively manage the development of new engineering products, the following has to be accomplished:

 (i) *Determine the strong and weak points of the company:* This can be accomplished only through the analysis of areas such as company reputation, facilities, financial capabilities, markets, executive talent, corporate image and board, and manpower quality and quantity.

 (ii) *Determine the goals of the company:* This is concerned with establishing the long range and overall goals of the firm. When determining the long range objectives, one has to establish company marketing objectives, manpower need, raw materials, prospects for business, financial condition, technical goals, etc.

 (iii) *Product Planning Function:* This is essentially concerned with the product planning function organization. According to Reference [6], for new engineering product organization, the guiding principles are as follows:

   (a) Take full advantage of the concept of inter-departmental product team.
   (b) New products are the responsibility of top management.
   (c) During all phases of product development, the marketing aspects must not be forgotten.
   (d) The process of new product must be managed.
   (e) A series of evaluations are involved with the new product development.
   (f) The function of new engineering product development must be established so that it is a top management activity.
   (g) The organization and control must be formed so that full consideration is given to the new product evolution stages.

12.2.4 Product Manager

The title of product manager is not as new as it may be thought to be. According to Reference [7], the origin of such title goes back as far as 1894 when the General Electric Company adopted it. However, generally the credit is given to Proctor and Gamble for introducing such a concept in 1928. This section briefly presents t͏he interfaces and responsibilities of a product manager.

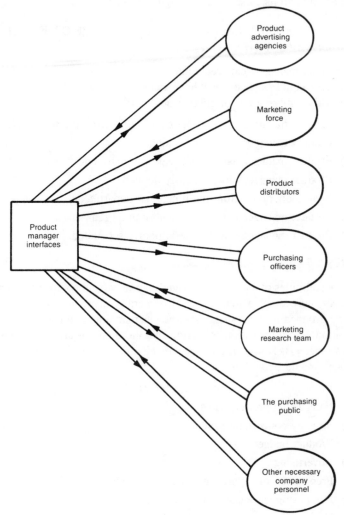

Figure 12.2.
Product manager's interfaces.

PRODUCT MANAGER INTERFACES

A product manager has to interface with various organizations to perform his duties effectively. Thus, according to Luck [8], the product manager interfaces are numerous as specified in Figure 12.2.

CAUSES FOR PRODUCT MANAGER FAILURE TO DEVELOP EFFECTIVE RELATIONSHIPS WITH INTERFACES

A product manager may have relationships with the interfaces but their effectiveness could be questionable in many cases. Therefore, following are some of the causes for poor effectiveness [3]:

   (i) Product manager receives poor assistance from the organization.

  (ii) Product manager has poor training.

 (iii) Product manager has to deal with too many interfaces.

 (iv) Product manager's job is poorly defined.

  (v) Poor cooperation from the functional departments.

 (vi) Product manager is occupied with unimportant tasks.

(vii) Product manager's work time is poorly scheduled.

DUTIES OF A PRODUCT MANAGER

Responsibilities of a product manager may vary from company to company. However, according to Reference [7], generally a product manager's duties may be grouped into classifications such as planning, control, co-ordination, evaluation, and information collection. All these groups are briefly described in Reference [7]. However, more specifically, the responsibilities of a product manager should be as follows:

    (i) Profit and loss associated with the product

   (ii) Forecasting product sales

  (iii) Strategy concerning product

  (iv) Product budget

   (v) Advertising

  (vi) Preparation of reports concerning product

 (vii) Purchasing associated with the product

(viii) New (related) product responsibility

  (ix) Direct control over product's technical development

   (x) Product quality control

  (xi) Interpretation of reports associated with the product

 (xii) Capital expenditure associated with the product

(xiii) Product pricing strategy

(xiv) Product related trouble shooting

12.2.5   Causes of Newly Developed Product Failures in the Market

There are various causes due to which the newly developed product fails to sell in the market. Some of these causes are as follows:

(i) *Poor quality:* The poor quality of the product is one of the important reasons for the poor sales of the product in the market.

(ii) *Poor Timing:* The newly developed product is introduced to the market at the wrong time.

(iii) *Wrong Market:* In this case the product is being marketed to an irrelevant market.

(iv) *Poor Design:* The design of the product is poor because of lack of care given at the design stage.

(v) *Product Price:* The product price range is wrong.

(vi) *Product Marketing:* The product is not being sold that well in the market because of poor market planning.

12.2.6   Hints to Avoid Product Failure

This section presents hints which will help to reduce the new product failure in the market [1]. Some of these hints are as follows:

(i) *Market Test:* It is always wise to test markets on a restricted basis before producing the product in mass quantity.

(ii) *Finance Study:* Economic and profitability studies by the finance department must be conducted when developing a new product.

(iii) *Feasibility Study I:* This must be conducted to obtain information on the feasibility of the product by the research and development department.

(iv) *Market Research Study:* This must be conducted by the marketing department.

(v) *Feasibility Study II:* This is to be conducted by the product manufacturing department.

(vi) *Team Effort:* Above all, the team approach among the marketing, finance, manufacturing and research departments is necessary right from the inception of the new product development program.

12.2.7   A List of Useful Information for Product Development

The information presented in this section will be very valuable when developing new engineering products. Thus, according to Petersen [9], information on the following would provide useful inputs for making decisions associated with new product development:

   (i) Time and cost to develop new product
  (ii) Manufacturing and development difficulties
 (iii) Prospects and profit in the long term
  (iv) Time required to break even
   (v) Time required to absorb product development costs
  (vi) Problems associated with patent
 (vii) Quantity of product units required
(viii) Capital requirement
  (ix) Estimate of product selling price
   (x) Profit margin and return on investment cost
  (xi) Foothold on new technology due to the development of product under consideration
 (xii) Materials related difficulties

12.3   PRODUCT COSTING

Whatever product an engineering company manufactures, it has to purchase some kind of raw or other materials from outside, to employ manpower to carry out the work, to pay for overheads, to know what the capital requirements are, etc. Therefore, all these items require cost estimations to make effective decisions. Furthermore, cost estimations play an important role in the future planning of a company. Without the proper cost estimations such plans may not be effective.

This section presents various aspects of product costing.

12.3.1   Reasons for Product Costing

There are various reasons for product costing. Some of them are as follows [10,11]:

   (i) To ascertain in the manufacture of a product, the most profitable material and procedure
  (ii) To determine the product price for the use in various purposes
 (iii) To establish the profitability of the product under consideration with respect to manufacture and market
  (iv) To establish the amount of money to be spent in equipment and other related items to produce a product
   (v) To determine if it is cheaper to fabricate the components or assemblies or procure them from outside agencies
  (vi) To study whether or not it is economical to modify the existing production facilities to manufacture a product

(vii) To provide input to the long-term company financial plans

(viii) To perform new products' feasibility analysis

(ix) To verify bids from outside agencies

(x) To measure the product manufacturing process efficiency

(xi) To provide assistance in controlling the costs of product manufacturing

Most of the above reasons are described in detail in Reference [11].

12.3.2 System Cost Estimation Procedure

Broadly speaking, there is no fixed procedure which can be followed to obtain a reliable system cost estimate. However, according to Reference [12], a general procedure which can usually be employed to estimate a system cost is shown in Figure 12.3. The steps of this approach are briefly described below.

(i) *Problem Definition:* This step is concerned with defining the system problem for which the cost is to be estimated. This can be accomplished properly only if there are close contacts between the cost and system analysts. Before the system cost estimation process

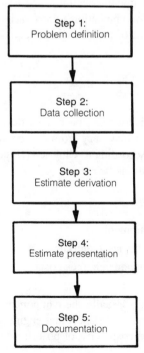

Figure 12.3.
A general procedure for estimating a system's cost.

can begin, there is a definite need for establishing cost ground rules and system description.

(ii) *Data Collection:* Data are the backbone of the cost study because the overall system cost estimate can be obtained only if the necessary data are available. Therefore, this step of the procedure is concerned with collecting various types of data which are vital to estimate system cost.

(iii) *Estimate Derivation:* Once the necessary data are available, the next logical step is to perform actual cost estimate calculations. Therefore, this step is concerned with deriving the estimate for costs.

(iv) *Estimate Presentation:* This is the fourth step of the system cost estimation procedure and is concerned with presenting the system cost estimates obtained from step (iii) to the users of these estimates. The estimates have to be presented in a way which can easily be utilized by the user to make necessary decisions. Therefore it is vital to establish early contacts between the cost analysts and the users to review jointly the presentation of the cost estimate results..

(v) *Documentation:* This is the final step of this procedure and is concerned with the documentation of the accomplished study. The end report of the documentation serves various purposes, for example, as a record of the study, a data source for future similar studies, etc.

12.3.3   Cost Estimation

The overall cost in a factory can be broken down into various elements. Such elemental breakdown may vary from one company to another or from one industry to another. According to Clugsten [11] cost can generally be sub-divided into the classifications below:

(i)   Cost of tooling

(ii)   Cost of overheads

(iii)   Cost of bought equipment

(iv)   Costs of raw and finished materials

(v)   Testing and direct labour charges costs

(vi)   Cost of pattern

(vii)   Cost of engineering — Usually the components of this cost are the costs of detail, layout and conceptual engineering, engineering associated overhead costs, cost of post-drawing issue modification, etc.

Generally in real-life situations, the capital investment cost and operating cost estimates are frequently used. These two items are separately discussed below.

ESTIMATION OF NEW PROJECT CAPITAL INVESTMENT COST

Estimating the fixed capital requirements cost is usually divided into three parts as follows [13]:

(i) *Cost of fixed investment, Type I:* This type of investment is depreciable and covers the cost of shipping and receiving facilities, transportation cost, and costs of equipment, building and services.

(ii) *Cost of fixed investment, Type II:* This type of investment is non-depreciable and includes the working capital and the cost of land.

(iii) *Cost of the amortized investment:* The components of this cost are the cost of changes (i.e., to make the product perform at the optimal point) after the product project is accomplished, cost of engineering and supervision, cost of research and development, etc.

ESTIMATION OF OPERATING COST

According to Reference [13], this is the cost of running a project. The components of this cost are as follows:

(i) Cost of general overheads

(ii) Cost of marketing

(iii) Cost of administrative activities

(iv) Operating costs, type I

(v) Operating costs, type II

The costs of general overheads include the costs of storage facilities, payroll overhead, cafeteria, etc. The components of marketing cost are the costs of advertising, marketing employees, shipping, etc. Administrative costs consist of items such as management personnel, clerical staff, legal activities and office maintenance. The type I operating costs cover all the direct operating costs. These costs are the variable costs, for example, costs of energy, maintenance, operating labor and raw materials.

Finally, the type II operating costs include essentially the indirect operating costs. These costs are the fixed costs, for example, rent, depreciation, interest, taxes and insurance.

12.3.4 Life Cycle Cost of a Product

The concept of life cycle costing is widely used in procuring the military hardware. The first usage of the term "life cycle costing" goes back to 1965 when a document entitled "Life Cycle Costing in Equipment Procurement" was published by the Logistic Management Institute, Washington, DC. The

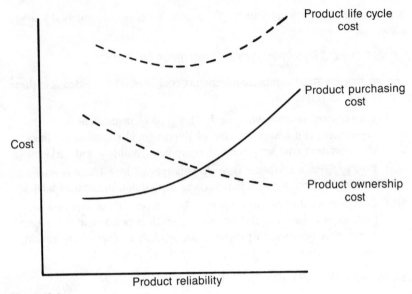

Figure 12.4.
Plots of product reliability against costs.

life cycle cost of a product is the total cost of that product over its entire life span. Basically, the life cycle cost of a product is composed of procurement and ownership costs. According to Reference [14], the cost of equipment ownership, depending on equipment type, varies from 10 to 100 times its procurement cost.

Therefore, when purchasing a new product the procurement management in many organizations today examines the life cycle cost of a product rather than just the initial procurement cost of that product. Figure 12.4 exhibits the relationship between the product reliability and costs.

There are various mathematical models used to calculate equipment life cycle cost [15]. One of these models uses the following equation to calculate the equipment life cycle cost.

$$c_\ell = \sum_{i=1}^{2} c_i \qquad (12.1)$$

where

c_ℓ is the equipment life cycle cost.
c_1 is the recurring cost of the equipment.
c_2 is the non-recurring cost of the equipment.

The equipment recurring cost covers the costs of maintenance, labour, inventory, support and operation. Similarly, the equipment non-recurring cost includes costs such as research and development, installation, support, training, reliability and maintainability, transportation, test equipment, management activities associated with life cycle costing, product qualification approval, and acquisition.

12.3.5   An Approach to Performing Cost-Effectiveness Analysis

This section presents a ten-step procedure for performing cost-effectiveness analysis [16]. These steps are described briefly below.

(i) Define the objectives of the system.

(ii) Outline the mission requirements pertinent to fulfilling system objectives.

(iii) To fulfill the missions, develop the concepts of alternative systems.

(iv) Develop a criterion to evaluate a system. This criterion must relate the capabilities of the system to the requirements of the mission.

(v) Choose either the fixed-effectiveness or fixed cost procedure. In the fixed-effectiveness the criterion used is the cost or resources required to obtain the specified effectiveness. Similarly, in the case of fixed cost, the criterion used is the effectiveness obtained with the utilization of specified resources..

(vi) Evaluate alternative systems capabilities.

(vii) Develop systems against criterion array.

(viii) Evaluate alternative systems advantages.

(ix) Conduct analysis of sensitivity.

(x) Document the study.

All the above ten steps are self-explanatory. Their detailed descriptions, however, are given in Reference [16].

12.3.6   Pricing of a New Product

The appropriate price of a new product is very important because it will influence people such as consumers, public policy makers, wholesalers, etc. Thus, the factors such as profit, units of product sales, and revenue are dependent on the product price. The following four factors play an important role in pricing a product:

(i) Competitive environment

(ii) Cost

(iii) Behaviour of the market

(iv) Demand for the product

In any situation, before the price of a product can be set, a company usually has its pricing objectives. Therefore, the price of a new product is set by taking into consideration the pricing objectives of the company. These objectives may vary from one product to another and from one company to another company. From Reference [17], some of the potential pricing objectives are as follows:

(i) To increase credibility of the firm

(ii) To maximize profits in the long term

(iii) To discourage new entrants in the market

(iv) To win the consumers' confidence as being a "fair" producer

(v) To avoid government interference

(vi) To maximize profits in the short term

(vii) To create interest in the new product

(viii) To discourage competitors from reducing their product prices

(ix) To assure that the middlemen loyalty is kept

(x) To bring the product to a "visibility" level

PRICING MODEL I

This model is concerned with the break-even analysis. This analysis is very simple and useful for investigating the effect on profits by changing the price and volume of the product sales units.

The term "break-even" pinpoints that point where the total cost to produce the product units is equal to the total revenue from the sales of those product units. From Reference [18], the equation for the break-even analysis is as follows:

$$s_p\, y = c_f + c_v \cdot y \qquad\qquad (12.2)$$

where

c_f is the fixed cost.

c_v is the variable cost per unit.

s_p is the product selling price for one unit.

y is the number of units to be sold for break-even.

The above equation can be utilized to calculate the break-even demand units for the product for different unit prices. This will help to decide the desirable unit price of the product.

EXAMPLE 12.1

An engineering company has estimated a fixed cost of $60,000 to develop an engineering product. Furthermore, the estimated variable cost of the product for one product unit is $4. Determine the quantity of units to be sold at the following prices to break even:

  (i) $5/unit

  (ii) $6/unit

  (iii) $4.5/unit

  (iv) $7/unit

Thus substituting the specified data into Equation (12.2) results in

$$y \cdot s_p = 60{,}000 + 4\,y \tag{12.3}$$

Solving the above equation for y, we get

$$y = \frac{60{,}000}{(s_p - 4)} \tag{12.4}$$

At $s_p = 5/unit, the number of units to be sold for break-even from Equation (12.4) is

$$y = \frac{60{,}000}{(5 - 4)} = 60{,}000 \text{ units}$$

Similarly for $s_p = 6/unit, $s_p = 4.5/unit, and $s_p = 7/unit, the number of units to be sold for break-even from Equation (12.4), is, respectively

$$y = \frac{60{,}000}{(6 - 4)} = 30{,}000 \text{ units}$$

$$y = \frac{60{,}000}{(4.5 - 4)} = 120{,}000 \text{ units}$$

and

$$y = \frac{60{,}000}{(7 - 4)} = 20{,}000 \text{ units}$$

Now with the aid of the above calculations, it is easier to make unit price decisions.

PRICING MODEL II

This model is used to determine the optimal price of a product unit [18,19]. The model makes use of price/demand relationship data. In this model it is assumed that the demand for a product follows the linear relationship as follows:

$$D = \alpha + \beta \cdot y \qquad (12.5)$$

where

D is the total demand (units) for the product.
y is the price of a single product unit.
β is a parameter (the slope of the straight line).
α is a parameter (given by the value of D when y is equal to zero).

Similarly, the total cost, c_t, of a product is given by

$$c_t = c_f + c_v D \qquad (12.6)$$

where

c_f is the product fixed cost.
c_v is the product variable cost per unit.

Thus total revenue, R_t, from the product is given by

$$R_t = \text{(price of a single unit)} \cdot \text{(Quantity of product sold)}$$
$$= y(\alpha + \beta \cdot y) \qquad (12.7)$$

Thus the profit, P, from Equations (12.6) and (12.7) is

$$P = R_t - c_t = y(\alpha + \beta \cdot y) - c_f - c_v \cdot D$$
$$P = y\,\alpha + \beta \cdot y^2 - c_f - c_v \cdot D \qquad (12.8)$$

Substituting for D from Equation (12.5) into Equation (12.8) yields:

$$P = y\alpha + \beta\,y^2 - c_f - c_v(\alpha + \beta\,y)$$
$$= y(\alpha - c_v \cdot \beta) + \beta\,y^2 - c_f - c_v \cdot \alpha \qquad (12.9)$$

Differentiating Equation (12.9) with respect to y yields

$$\frac{dP}{dy} = (\alpha - c_v \cdot \beta) + 2\,\beta y \qquad (12.10)$$

Setting the left-hand side of Equation (12.10) equal to zero and solving for y leads to:

$$y^* = \frac{c_v}{2} - \frac{\alpha}{2\beta} \qquad (12.11)$$

where y^* is the optimal price of a product unit.

The quantity of product units that will be sold at the price of y^* is obtained by substituting Equation (12.11) into Equation (12.5) as follows:

$$D = \alpha + \beta \left(\frac{c_v}{2} - \frac{\alpha}{2\beta} \right)$$
$$= \frac{\alpha + \beta c_v}{2} \qquad (12.12)$$

EXAMPLE 12.2

A manufacturer produces a certain engineering product whose fixed and variable costs are $140,000 and $16 (per unit), respectively. After testing the market, the manufacturing company collects the following price/demand data:

Estimated product price (per unit)	Estimated demand for the product (units)
$100	500
$ 80	1,000
$ 60	1,500
$ 40	2,000
$ 20	2,500

Determine the optimum price of a single product unit.

The plot of the price/demand data indicates the straight line relationship. The straight line intercepts the vertical axis (product demand axis) at 3000 units and the slope of the straight line is equal to –25.

Thus the values of α and β of Equation (12.11) are 3,000 and –25, respectively. Furthermore, $c_v = \$16$ per unit.

Substituting the above specified data into Equation (12.11) leads to

$$y^* = \frac{16}{2} - \frac{3,000}{2(-25)}$$
$$= \$68$$

Thus the optimal price of a product unit is $68.

12.3.7   Maintenance Cost Formulas

This section presents three proposed formulas in the literature to calculate maintenance costs of various types of plants. These formulas are presented below [20–22].

FORMULA I

This formula is proposed to estimate the maintenance cost of a large coke plant. The formula is expressed as follows:

$$C_{mc} = 0.004\beta - 83{,}091 \tag{12.13}$$

$$\beta \equiv \sum_{i=1}^{k} \alpha_i t_i \tag{12.14}$$

where

C_{mc} is the coke plant yearly maintenance cost in dollars.
α_i is the ith production unit investment.
t_i is the ith production unit time in years from its installation.
k is the number of production units maintained.

FORMULA II

This is another formula proposed to estimate the maintenance cost of a large cement plant. The formula is expressed as follows:

$$C_{cpm} = 0.011\beta - 300{,}000 \tag{12.15}$$

$$\beta \equiv \sum_{i=1}^{k} \alpha_i t_i \tag{12.16}$$

where C_{cpm} is the cement plant yearly maintenance cost in dollars.

FORMULA III

This formula is proposed to estimate the maintenance cost of a large pulp and paper plant. The formula equation is expressed as follows:

$$C_{pm} = 0.009\beta + \sigma \tag{12.17}$$

$$\beta \equiv \sum_{i\,=\,1}^{k} \alpha_i t_i \qquad (12.18)$$

$$\sigma = 149,000 \qquad (12.19)$$

where C_{pm} is the pulp and paper plant yearly maintenance cost in dollars.

We should note here that before using Equations (12.13) – (12.19) in real life situations one should consult References [20–22].

12.4 SUMMARY

This chapter briefly describes the two important topics of engineering product developing and costing. Thus the chapter is divided into two parts. The first part is concerned with product developing. Under this subject seven important topics related to product developing are briefly described. These topics cover the areas such as reasons for developing new engineering products, a procedure to develop new products, product manager, causes of new product failures and hints to avoid product failures in the market.

The second part of the chapter concentrates on product costing. This section begins by listing eleven reasons for product costing and goes on to describe a procedure for estimating a system cost. The system cost estimation procedure is composed of five steps. The next topic of this section explains the components of capital investment cost and operating cost. The concept of life cycle costing is briefly discussed. The other two important topics discussed in this section are cost-effectiveness analysis and pricing of a new product. The steps of a cost-effectiveness procedure are presented and two mathematical models useful in new product price setting are described. Lastly, this section presents three formulas for estimating maintenance cost of three different types of plants.

Source references for the material presented in the chapter are given in the reference section of this chapter.

12.5 EXERCISES

1. Explain why there is the need for an engineering company to develop new engineering products.
2. Describe the following terms:
   (i) Capital investment
   (ii) Product pricing

    (iii) Cost-effectiveness of a system
    (iv) Life cycle cost

3. Describe the important factors to be considered in the product planning function organization.

4. What are the interfaces of a product manager?

5. What are the functions of a product manager?

6. Describe the reasons for product failures in the market.

7. Why is the costing of a new engineering product necessary?

8. What are the main reasons for establishing the price of an engineering product?

9. Compare the responsibilities of a product development manager with that of a product costing manager.

10. After testing a newly developed engineering product in the market, an engineering company established the following price/demand relationship:

$$d = 2000 - 15\,x$$

where

d is the demand (in units) for the new product.
x is the price of a single unit.

The fixed and variable costs associated with this product are $80,000 and $12 (per unit), respectively. Calculate the optimum price of a single product unit.

12.6 REFERENCES

1. Manning, C. F. "How New Products Can Increase Profits," in *Management Guide for Engineers and Technical Administrators.* N. P. Chironis, ed. New York:McGraw-Hill Book Company (1969).

2. Federal Trade Commission. "Permissible Period of Time During Which New Product May be Described as New," *Advisory Opinion Digest,* 120 (April 15, 1967).

3. Spitz, A. E., ed. *Product Planning.* Princeton:Auerbach Publishers (1972).

4. Buzzell, R. D. and R. Nourse. *Product Innovation in Food Processing: 1954–1964.* Boston:Division of Research, Harvard Business School (1967).

5. Hise, R. T. *Product/Service Strategy.* New York:Petrocelli/Charter (1977).

6. Karger, D. W. and R. G. Murdick. *Managing Engineering and Research.* New York:Industrial Press Inc. (1969).

7. Baker, M. J. and R. McTavish. *Product Policy and Management.* London:The Macmillan Press Limited (1976).

8. Luck, D. J. "Interfaces of a Product Manager," *Journal of Marketing,* 33 (Oct. 1969).

9. Petersen, J. W. "A Workable Approach to New-Product Planning," in *Management Guide for Engineers and Technical Administrators*. N. P. Chironis, ed. New York:McGraw-Hill Book Company (1969).

10. Doyle, L. E. "How to Estimate Costs of New Products," in *Management Guide for Engineers and Technical Administrators*. N. P. Chironis, ed. New York:McGraw-Hill Book Company (1969).

11. Clugston, R. *Estimating Manufacturing Costs*. 89 Franklin Street, Boston, MA 02110:Cahners Books.

12. Goldman, T. A., ed. *Cost-Effectiveness Analysis: New Approaches in Decision-Making*. New York:Frederick A. Praeger, Publishers (1971).

13. Kasner, E. *Essentials of Engineering Economics*. New York:McGraw-Hill Book Company (1979).

14. Ryan, W. J. "Procurement Views of Life Cycle Costing," *Proceedings of the Annual Symposium on Reliability*. pp. 164–168 (1968).

15. Dhillon, B. S. *Reliability Engineering in Systems Design and Operation*. New York:Van Nostrand Reinhold Company (1982).

16. Kazanowski, A. D. *A Standardized Approach to Cost-Effectiveness Evaluations in Cost-Effectiveness: The Economic Evaluation of Engineered Systems*. New York:John Wiley & Sons, Inc. (1968).

17. Oxenfeldt, A. R. "A Decision-Making Structure for Price Decisions," *Journal of Marketing*, 50 (1973).

18. Hisrich, R. D. and M. P. Peters. *Marketing a New Product: Its Planning, Development and Control*. Menlo Park, CA:The Benjamin/Cummings Publishing Company, Inc. (1978).

19. Harper, D. V. *Price, Policy and Procedure*. New York:Harcourt, Brace and World, pp. 143–148 (1966).

20. Hackney, J. W. *Control and Management of Capital Projects*. New York:John Wiley & Sons, Inc. (1965).

21. Robbins, M. D., ed. "Maintenance Cost Estimation, Is This a Better Tool for You?" *Chemical Engineering*, 140–142 (Feb. 9, 1959).

22. Robbins, M. D., ed. "Predicting Your Maintenance Costs," *Chemical Engineering*, 172–178 (July 13, 1959).

CHAPTER 13

Management of Engineering Design

13.1 INTRODUCTION

Each day newly designed engineering equipment is put into service and old equipment is discarded. Each new engineering product sold in the market has to be designed within general guidelines such as (i) satisfying the customer's need; (ii) keeping the selling price of the product within the customer's expectation, etc. From the manufacturer's and customer's point of view the best designed product is that which brings them the most profit. Thus, the design of the product under development has to be managed in such a way that all the necessary design guidelines and other associated factors are fully satisfied.

According to Reference [1], the engineering design may be described as an iterative decision-making activity for developing the plans by which resources are transformed, to fulfill human requirements, into systems, sub-systems or components. This chapter briefly discusses various aspects of engineering design.

13.2 DEMAND FOR DESIGN

The design process can begin only if there is a demand. According to Reference [2], there are basically four ways, as shown in Figure 13.1, in which the demand for design may arise. These are described as follows [2].

DEMAND FROM USER

Some of the reasons for the user demand are:

 (i) Potential demand materialization
 (ii) Increase in standard of living
 (iii) Competition
 (iv) Rise in labour costs, requiring more and better products
 (v) Need for a better product at the same price
 (vi) Improvement in public taste
 (vii) Ownership pride

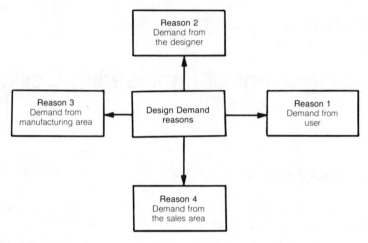

Figure 13.1.
Reasons for design demand.

(viii) Realization by the user that his specific need required from the existing product can be fulfilled only by the new product

DEMAND FROM THE DESIGNER

This may arise due to the following factors:

(i) To enhance his professional status.
(ii) Due to past product design experience, he can visualize room for further improvement in another product design.
(iii) Creating new products is his hobby.
(iv) To gain financial reward.

DEMAND FROM THE MANUFACTURING AREAS

The following are the contributory reasons for the demand from the manufacturing areas:

(i) Manpower (labour) quality
(ii) Plant quality
(iii) Raising profit
(iv) Pressure due to economics
(v) Availability of modern production techniques
(vi) Manpower (labour) availability
(vii) Plant availability
(viii) To decrease production difficulties

DEMAND FROM THE SALES AREAS

Some of the contributing factors for the demand from the sales area are as follows:

(i) To raise company profit
(ii) Strategic importance
(iii) To increase sales volume
(iv) Competition
(v) To gain more prestige
(vi) Political reasons

13.3 TYPES OF DESIGN WORK

According to Reference [3], there are three types of design work. These are as follows:

(i) Adaptive design work
(ii) Developmental design work
(iii) New design work

These three are briefly described as follows.

ADAPTIVE DESIGN WORK

This design work, in very many cases, is simply concerned with adaptation of already existing designs. In some situations the designer may simply be making some minor modifications. Usually with simple technical training, a designer can solve such design activity problems. However, a person who usually works from already developed designs may not appreciate fully the meanings of the design until he is required to input original thoughts to a simple or complex task. Normally, the fresh designers go through the route of the adaptive design work.

When it is necessary to modify already existing designs, for example, making a switch from one method of manufacture to another, then it certainly requires a better degree of designing ability.

DEVELOPMENTAL DESIGN WORK

This type of design work requires more experience and expertise in comparison to the adaptive design work. In the developmental design work, the designer initiates the developmental design from the already existing design. However, the final designed product may be significantly different from the product at the starting stage.

NEW DESIGN WORK

This is concerned with venturing into new design fields. Therefore, it requires a very high degree of competence from the designers. Only a handful of designers end up doing this type of design work.

13.4   EIGHT-STEP PROCEDURE USED IN DESIGNING

This section describes a useful eight-step procedure used in designing [4]. These steps, shown in Figure 13.2, are concerned with analysing need, defining problems, finding alternative solutions to the problem, analysing solutions

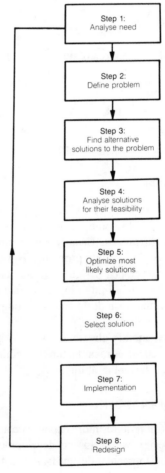

Figure 13.2.
An eight-step approach for designing.

for their feasibility, optimizing the most likely solutions, and selecting, implementing, and redesigning solutions.

The first several steps given in Figure 13.2 are described below.

STEP 1: ANALYSE NEED

This step is concerned with analysing the following needs of the user:

(i) Social and cultural needs

(ii) Physiological needs

(iii) Technical needs

(iv) Psychological needs

In order to produce the acceptable design for the user, these needs have to be fully satisfied. These needs are described in more detail in Reference [5].

STEP 2: DEFINE THE PROBLEM

In relation to the needs, here one is concerned with defining the objectives, major restraints, variables, boundary, inputs and outputs. More clearly, at this stage the following questions are asked:

(i) What is the design problem under consideration?

(ii) What are the objectives to be satisfied by the design under consideration?

(iii) What are the social consequences associated with the design?

(iv) Is the design problem too big?

(v) What are the constraints associated with the design?

(vi) Is there any specific difficulty?

(vii) What should be contained in the problem definition?

Step 2 is described in detail in Reference [6].

STEP 3: FIND ALTERNATIVE SOLUTIONS TO THE PROBLEM

This step is concerned with finding a range of alternative design solutions by utilizing past experience and creative techniques. The brainstorming method is one of the creative techniques. This step is described in detail in Reference [7].

STEP 4: ANALYSE SOLUTIONS FOR THEIR FEASIBILITY

Once a number of design solutions are obtained, the next step is to thoroughly examine these alternative solutions for their physical, financial, social and cultural, and economic feasibility. The physical, financial and economic feasibility are discussed briefly as follows:

(i) *Physical Feasibility:* Any new design will be physically feasible only if it obeys the laws of nature. Furthermore, if each alternative solution is treated as a system then one may ask the following questions:

(a) What are the difficulties which may appear at the subsystem interconnections?

(b) Will the performance of integrated subsystems be in line with the requirement for the overall system?

(c) Will the system transform the specified inputs to the specified outputs?

(ii) *Financial Feasibility:* This is concerned with the financial aspects of the design. More clearly, here one examines whether the capital to finance the entire project, from start until the investment is recovered with some profit, is enough.

(iii) *Economic Feasibility:* In simple terms, here one is concerned with the economic aspect of the design solutions. All in all, any design solution will be attractive only if it is economically acceptable. Furthermore, the following questions are also asked during the economic feasibility analysis:

(a) Will the proposed design solution enhance the firm's reputation?

(b) Can it withstand a normal overload?

(c) Will it function normally during its specified warranty period?

(d) Will it fulfill user requirements reliably?

STEP 5: OPTIMIZE THE MOST LIKELY SOLUTIONS

After accomplishing the above step, we end up with some closely competing alternative solutions. More clearly, these alternatives seem to have met the physical, financial and economic feasibility requirements. Thus, the next step is to establish which is the best alternative design solution by optimizing all the design solutions which have passed through the feasibility investigation. There are various procedures used in optimizing the engineering design, for example, calculus, linear programming, a graphical model, a physical model, etc. [9]. A simple optimization problem [10] is demonstrated in the following example.

EXAMPLE 13.1

A person wishes to enclose a plain surface rectangular area with a fence. Furthermore, he wishes to accomplish his objective with a minimum amount of fencing. Assume that the field length, α_f, to width β_f, ratio is subject to selection. Develop an expression for the optimum length of fence if the field area, a_f, is defined.

Thus, the ratio between α_f and β_f is the variable, θ, given by

$$\theta = \beta_f/\alpha_f \qquad (13.1)$$

Therefore,

$$\beta_f = \theta \, \alpha_f \tag{13.2}$$

The field area, a_f, is given by

$$a_f = \beta_f \cdot \alpha_f \tag{13.3}$$

Substituting Equation (13.2) into Equation (13.3) yields

$$a_f = \theta \, (\alpha_f)^2 \tag{13.4}$$

From the above equation we get

$$\alpha_f = \left(\frac{a_f}{\theta} \right)^{1/2} \tag{13.5}$$

The total length of the fence is given by

$$L_t = 2(\alpha_f + \beta_f) \tag{13.6}$$

Incorporating Equation (13.2) into Equation (13.6) results in

$$L_t = 2(\alpha_f + \theta \, \alpha_f)$$
$$= 2 \, \alpha_f(1 + \theta) \tag{13.7}$$

Substituting Equation (13.5) into Equation (13.7) yields

$$L_t = 2 \left(\frac{a_f}{\theta} \right)^{1/2} (1 + \theta) \tag{13.8}$$

Rewriting the above equation in slightly different form results in

$$L_t = 2(a_f)^{1/2} \left(\theta^{1/2} + \frac{1}{\theta^{1/2}} \right) \tag{13.9}$$

Differentiating Equation (13.9) with respect to θ leads to

$$\frac{dL_t}{d\theta} = 2(a_f)^{1/2} \left(\frac{1}{2} \theta^{-1/2} - \frac{1}{2} \theta^{-3/2} \right)$$
$$= (a_f)^{1/2}\{\theta^{-1/2} - \theta^{-3/2}\} \tag{13.10}$$

To determine the optimum value of L_t we set Equation (13.10) equal to zero:

$$(a_f)^{1/2}(\theta^{-1/2} - \theta^{-3/2}) = 0 \qquad (13.11)$$

or

$$(\theta^{-1/2} - \theta^{-3/2}) = 0 \qquad (13.12)$$

At $\theta = 1$, the above equation will be satisfied. Thus, the optimum value expression for L_t, from Equation (13.9), is

$$L_t^* = 2(a_f)^{1/2} \left[(1)^{1/2} + \frac{1}{(1)^{1/2}} \right]$$

$$\qquad (13.13)$$

$$L_t^* = 4(a_f)^{1/2}$$

where $a_f = \beta_f^2$ or α_f^2.

Thus in the concluding remarks it is added that the shape of the field with specified area has to be square (i.e., $\beta_f = \alpha_f$) in order to minimize the fencing cost.

EXAMPLE 13.2

A company has a large plot of land. The organization wishes to fence 90,000 square yards of land for its day-to-day use. The overall cost, c_f, of fencing is $50 per yard. Calculate the optimum total cost to the company of fencing.

From the earlier analysis it is proven that the field shape has to be square. Thus substituting the specified data for $a_f = 90,000$ square yards into Equation (13.13) yields

$$L_t^* = 4(90,000)^{1/2} = 1200 \text{ yards}$$

Thus, the minimum length of fence to cover the 90,000 square yards is 1200 yards. The total minimum cost of fencing, c_t, is

$c_t =$ (Optimum total length of the fence) $\cdot$ (Per unit overall cost of the fence)

$= (L_t^*) \cdot (c_f)$

$= (1200)(50)$

$= \$60,000$

The optimum total cost of fencing will be $60,000.

STEP 6: SELECT SOLUTION

With the aid of established design criterion, select an optimized design solution. The design criterion means established criterion for judging and evaluating alternative design solutions. The topic of establishing design criterion is described in Reference [11].

STEP 7: IMPLEMENTATION

This step is concerned with developing instructions, specification, and prototypes which can easily be understood by those who will use them.

STEP 8: REDESIGN

This step is self-explanatory, and after this step the cycle repeats as shown in Figure 13.2.

13.5 DESIGN INFORMATION SOURCES

In engineering design, generally various types of information are needed. According to Reference [2], this information can be divided into the following categories:

   (i) Standards data
  (ii) Marketing, costs and commercially oriented information
 (iii) Product usage information
  (iv) Design data
   (v) Basic engineering
  (vi) Administration approaches
 (vii) Methods of working

The following are the main sources which can be used to obtain information:

   (i) Product users, market research
  (ii) Journals, magazines, conference proceedings and technical reports
 (iii) Libraries, professional bodies, industrial organizations, academic institutions and other bodies
  (iv) Conferences, meetings, seminars, and courses
   (v) Organization's files, personal contacts and notes
  (vi) Trade catalogues, exhibitions and suppliers
 (vii) Results of experiments
(viii) Standards

13.6   DESIGN SPECIFICATION

When designing engineering products, the design specification plays a very important role. Therefore, the preparation of design specification requires special care because various information regarding the product to be designed is transmitted through the specifications. When writing a specification, one should make sure that

(i)   It is concise and clear.

(ii) It is flexible, so that further improvements are easily incorporated.

(iii) It is reasonable for specified tolerances.

(iv) It is accurate.

(v)   It is complete.

A good engineering design specification should incorporate information on items briefly presented in Figure 13.3. Some of these items are briefly described below [2].

(i)   *Operation related constraints:* These include product reliability, useful life, operating and maintenance procedures, power supplies, etc.

(ii) *Functional requirements:* These are performance, unacceptable failure modes, useful life, acceptable tolerances, etc.

(iii) *Manufacture related constraints:* These constraints are availability of development facilities, manpower and manufacturing processes; allowable cost of manufacturing, etc.

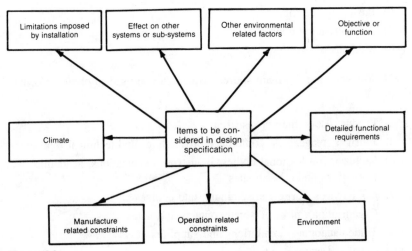

Figure 13.3.
Design specification areas.

(iv) *Environment:* This category is concerned with acceleration, contaminants, ambient temperature and pressure, and vibrations.

The remaining specification areas are self-explanatory and are described in detail in Reference [12].

13.7 MANAGEMENT EXPECTATIONS FROM AN ENGINEERING DESIGN DEPARTMENT AND THE POINTS ASSOCIATED WITH DESIGN REQUIRING DECISION

A company usually expects the following from its engineering design department:

(i) The design is sellable.

(ii) The design is producible.

(iii) The design is reliable.

(iv) The design is profitable.

(v) The design meets completion date.

(vi) The design is creative.

The design will be sellable in the market only if it meets requirements such as social, cultural, psychological, physiological and technical needs. In the design producibility aspect, the designer and the department have to make sure that the producibility requirements such as skilled manpower, availability of required technology, facilities, etc., are within reach.

The reliability factor is very important in the newly designed product. Therefore, the designer has to take into consideration that the designed product will function normally as specified when put to use. Ultimately, the companies will appreciate only those product designs which are going to be profitable ventures. Therefore, the design must go through a rigorous economic analysis.

Another important expectation from the engineering design department is that the design completion and other associated dates are met according to plans. Generally, in the competitive environment, the design can be profitable only if it can be completed on time.

Finally, the expectation of creative design is another area to which the top level company management looks from the design department. A creative design may open up new markets for the product if it enhances the product performance, appeals to users, reduces costs, and so on.

13.7.1 Points Associated with Product Design Requiring Decision

According to Reference [2], the following points concerning product design have to be decided:

(i) Liaison with suppliers, responsibility

(ii) Extent of responsibility for test requirements, appearance design, production and environment requirements

(iii) Division of responsibility, between various technical areas, for a particular product, for example, between electrical, mechanical or instrumentation design

(iv) Documentation requirement

(v) Project priority, budget costs evaluation and costs responsibility

(vi) Project responsibility

(vii) Type of design proposals required and contractual arrangement responsibility

(viii) Necessary approvals

(ix) Amount of research and investigation required

(x) Work breakdown

13.8   ATTRIBUTES OF AN ENGINEERING DESIGNER

Just like a manager, the designer of an engineering product has to possess certain qualities in order to perform his assigned task effectively. According to Reference [3], an engineering designer must have qualities such as follow:

(i) *Willpower:* In designing engineering products, a designer may be faced from time to time with up and down situations, objections and oppositions from others, etc. In the face of these situations, only those designers who have strong willpower will be successful.

(ii) *Ability to Remember:* In real life situations, a designer has to remember various facts and figures concerning design, his past experiences with the design work, etc. Therefore, a good memory plays an important role in the designer's future design work.

(iii) *Visual Capacity:* A designer must have the ability to visualize interpretations, sections, simple basic forms and their combinations. Static forces, and thermal, dynamic and electrical phenomena are typical examples of where the engineering designer's visual capacity can be utilized.

(iv) *Creativeness:* Creativity or inventiveness plays an important role in design work. For example, a designer with creative ability may design a product which has better performance, lower cost, etc. Therefore, this product is likely to be more profitable to the company.

(v) *Speaking and Writing Skills:* These two are the other two important attributes of a designer. In day-to-day life, a designer may run into situations where he has to convince others of the goodness of his proposed design. To handle this type of situation, a designer must have good speaking and writing ability.

(vi) *Capacity to Integrate:* This is another requirement which is expected from a designer. It is proven that a designer with integrating capacity may be able to create something new by properly combining already existing inventions or ideas. Therefore, one can say that the capacities to integrate and visualize are the two most important components of a designer's creative imagination.

(vii) *The Ability to Concentrate and Think Logically:* Design work demands a high level of concentration from a designer. Therefore, a designer who is going to be successful in his field has to have such an ability. Furthermore, a designer must be able to think logically so that he can distinguish essentials from non-essentials, determine correctly the cause and effect interrelationships, etc.

(viii) *Other Qualities of a Designer:* In addition to qualities (i)–(vii), an engineering designer must also possess integrity, personality, sense of responsibility, conscientiousness and perseverance.

13.9 MANAGEMENT OF DESIGN REVIEWS

Various reviews are conducted during the design phase of an engineering product to make sure that the design work is progressing according to the design specification, plans, etc. According to Reference [13], in general there are four types of design review. However, usually the review type is dependent upon the status of the design. The four types of design reviews are as follows:

(i) Concept phase design review

(ii) Pre-release design review

(iii) Large-quantity release design review

(iv) Miscellaneous design review

The review, which is conducted before the release of engineering drawings, helps to do the following [14]:

(i) To determine the design from the mechanical, thermal and electrical aspects

(ii) To identify causes which may down-grade the product reliability

(iii) To make certain the effort put into quality control will be effective

(iv) To make sure that the standard or preferred parts, and preferred circuitry are used as much as possible

(v) To assure that the product is designed according to the specification requirements

(vi) To provide assurance that the human factors aspects have received necessary attention

(vii) To make sure that the interchangeability of similar parts, circuits, sub-systems, etc., have been given the deserved consideration in the design

13.9.1   Design Review Team

In simple terms, the main objective of the team is to review product design so that the various areas of design as specified earlier in this section are evaluated.

Usually, the design review team members are classified into the following three categories:

(i) Technical personnel belonging to those departments or groups which are directly associated with the product development

(ii) Design specialists belonging to those groups which are not directly involved with the product development

(iii) Customer representatives (if applicable)

According to Reference [14], to obtain effective performance results, the design review team membership should be confined to twenty persons. Usually, the following technical specialists should be members of the design review group:

(i) Chairman of the review team

(ii) Engineer or engineers involved in designing the product

(iii) Quality control and reliability engineers or managers

(iv) Tooling and manufacturing engineers

(v) Field and material engineers

(vi) Engineer responsible for packaging and shipping

(vii) Design engineers who are not concerned with product under design

(viii) Procurement and customer representatives

The responsibilities of all these personnel are briefly described in Reference [14].

13.9.2   Design Review Team Chairman

This person chairs the meetings of the design review board. Therefore, careful consideration must be given when selecting a person for such a position. Factors such as a person's position in the organization, technical competence, and personality play an important role when appointing the chairman of the review board. Furthermore, this person must not be the member of any of the support groups or design staff. Usually, the review board chairman performs the following tasks:

   (i) He chairs the design review board meeting.

  (ii) He develops the procedure for choosing necessary items for design review and the nature or type of the forthcoming review.

 (iii) He schedules reviews according to the design progress of the items needing review.

 (iv) He coordinates with and assists the involved bodies in preparation of necessary data and circulation of necessary documents, such as drawings, data and agenda concerned with the review, to appropriate groups or departments.

  (v) He sends the minutes of the review meeting to the appropriate people or groups and directs the necessary follow-up actions after the review.

13.9.3   Areas of Design Review Questions

This section briefly discusses the general areas in which questions are usually asked to the (major) engineering product designer(s). These areas are as follows [13]:

   (i) Product reliability, maintainability and safety

  (ii) Human factors

 (iii) Finishing and drafting

 (iv) Value engineering and reproducibility

  (v) Adherence to specifications and standardization

Reliability is an important factor in product design. Therefore, in order to make sure that the product is reliable in field use, several questions regarding reliability are asked during the design review. Examples of such questions are as follows:

   (i) Is the quantitative reliability specified (if any), and is the design specification satisfied?

  (ii) What were the assumptions to which the product reliability evaluation was subjected?

Maintainability of engineering equipment in the field is another concern to the design review team. Thus, the following questions may be asked during the design review.

   (i) Will the product meet the specified downtime?

  (ii) Was any consideration given to the unit replacement maintenance?

Safety is another area about which many questions are asked, because, for example, the design may be reliable and maintainable according to the specified requirements, but it may be unsafe for human beings.

A good design must take into consideration the human factor. A design is good only if it fits well to the people who will operate or use the designed product. Therefore, during the design review many questions are asked on areas concerned with humans, such as logical arrangement of controls, meters, lights glare, etc.

Two other areas which are also investigated during the review are finishing and drafting. Questions on finishing may be concerned with paint finish, weld-laps, etc. Similarly, questions on drafting may deal with dimension groupings, datum lines, tolerances, etc.

The area of value engineering is also probed during the review of product design. A considerable financial improvement may be made by asking the right questions in the design review meeting on value engineering.

In regard to reproducibility, a workable product model may be constructed by the model maker, but will it be possible to do the same thing economically in the production shop? Therefore, various questions are asked on reproducibility.

Finally, examples of questions regarding adherence to specifications and standardization, respectively, are as follows:

(i) Is the correct color used in finishing of the product?

(ii) Why are non-standard components used in the product design when the standard ones are readily available?

13.10 SUMMARY

This chapter briefly summarizes the various areas of management of engineering design. These areas are as follows:

(i) Demand for design

(ii) Types of design work

(iii) Eight-step procedure used in designing

(iv) Design information sources

(v) Design specification

(vi) Management expectations from an engineering design department and the points associated with design requiring decision

(vii) Attributes of an engineering designer

(viii) Management of design reviews

The demand for design may originate from sources such as user, designer, and manufacturing and sales areas. Therefore, these four items are briefly discussed in the chapter. The next topic discussed in the chapter is the types of design work: adaptive design, developmental design and new design.

The third topic of the chapter is the procedure used in designing. This procedure is composed of eight steps: analysing need, defining problems, finding alternative solutions to the problem, analysing solutions for their feasibility, optimizing the most likely solutions, selecting solution, implementing, and redesigning. Each step is briefly discussed.

Design information sources is another topic which is dealt with in the chapter. Under this topic various sources of information are listed.

The chapter goes on to describe the design specification. Eight major items concerning the specification are briefly discussed.

The next topic of the chapter is the management expectations from an engineering design department. Therefore, this section briefly presents the six management expectations. In the next section, the chapter goes on to describe the qualities of an engineering designer.

The final topic of the chapter is the management of design reviews. This topic is described in some depth. The major areas discussed are the design review team, its chairman, and design review questions.

Sources for the material presented in the chapter are listed in the reference section.

13.11 EXERCISES

1. What are the reasons, which originate from the designer himself, for demand for the new product design?
2. Write an essay on the principle of design feasibility.
3. List the sources which are used to obtain design related information.
4. Write a design specification for an electric switch.
5. What are the functions of an engineering design department?
6. What are the similarities between the qualities of an engineering manager and an engineering designer?
7. Discuss the duties of the design review board chairman.
8. Who are the people who participate in a major engineering product design review? Describe the functions of each participant.
9. Reliability and maintainability are two areas about which questions may be asked during a design review. Write down at least ten questions on both these topics which can be asked during a design review meeting.
10. Describe the objectives of a design review.
11. Write down the advantages and disadvantages of a product design review.

13.12 REFERENCES

1. Woodson, T. T. *Introduction to Engineering Design.* New York:McGraw-Hill Book Company (1966).

2. Cain, W. D. *Engineering Product Design.* John Wiley & Sons, Inc. (1969).

3. Matousek, R. *Engineering Design.* New York:John Wiley & Sons, Inc. (1963).

4. Love, S. F. "Design Methodology," *Design Engineering,* 30–32 (April 1969).

5. Love, S. F. "Finding Out What is Really Wanted: The Principle of User Needs," *Design Engineering,* 41–43 (March 1977).

6. Love, S. F. "Design Methodology: Defining the Problem," *Design Engineering,* 69–72 (July 1969).

7. Love, S. F. "Design Methodology: Creating New Design Solutions," *Design Engineering,* 66–85 (October 1969).

8. Love, S. F. "Design Methodology: Feasibility of Design Solutions," *Design Engineering,* 60–62 (November 1969).

9. Love, S. F. "Design Methodology: Optimizing the Design," *Design Engineering,* 42–45 (December 1969).

10. Wilson, W. E. *Concepts of Engineering System Design.* New York:McGraw-Hill Book Company (1965).

11. Love, S. F. "Design Methodology: Establishing Design Criteria," *Design Engineering,* 42–44 (August 1969).

12. Leech, D. J. *Management of Engineering Design.* New York:John Wiley & Sons, Inc. (1972).

13. Simonton, D. P. "Way a Design-Review Committee Pays Off Dividends," in *Management Guide for Engineers and Technical Administrators.* N. P. Chironis, ed. New York:McGraw-Hill Book Company (1969).

14. "Engineering Design Handbook: Development Guide for Reliability, Part Two: Design for Reliability," Published by Headquarters, U.S. Army Material Command, Available from the National Technical Information Service, Springfield, VA (January 1976).

CHAPTER 14

Management of Engineering Drawings

14.1 INTRODUCTION

The history of technical drawings may be traced back to 4,000 B.C. when the Chaldean engineer Gudea engraved upon a stone tablet the plan view of a fortress [1]. However, the written evidence of use of technical drawings can be traced back to only 30 B.C. which is the date associated with a treatise on architecture written by the Roman architect Vitruvius.

It appears that the first book published in the United States on the subject of technical drawing was in 1849. The book was entitled *Geometrical Drawing* and was published by William Minifie.

Nowadays in the design, development and operation of engineering systems, engineering drawings play an important role. Without the use of engineering drawings, it may not be possible to design new engineering products.

In the development of each engineering system a number of engineering drawings are required. As the number of engineering drawings increases, so does the problem of their management. Therefore, the management of engineering drawings is very important in order to produce satisfactory engineering products. For example, due to lack of proper established channels to control changes in engineering drawings, a company may end up with wrongly manufactured products.

This chapter briefly describes the various important aspects of engineering drawings and their management.

14.2 TYPES OF TECHNICAL ILLUSTRATIONS AND DRAWINGS

This section briefly discusses various types of technical illustrations and drawings. According to Reference [2], there are five important types of technical illustrations. These are as follows:

(i) *Type I Illustrations:* These are known as axonometric projections, which are basically parallel-plane drawings depicting three dimensions.

(ii) *Type II Illustrations:* Diagrams and charts fall into this category.

257

Examples of diagrams and charts are wiring diagrams, block diagrams, schematics, graphs, flow charts, etc.

(iii) *Type III Illustrations:* These are known as perspectives. The perspectives are three-dimensional drawings. However, their picture planes converge to a point where they vanish.

(iv) *Type IV Illustrations:* This category includes orthographic drawings which are simply two-dimensional engineering projects.

(v) *Type V Illustrations:* These are essentially photographs or continuous-tone camera pictures. However, these are modified with the aid of art techniques to illustrate technical information.

The following types of drawings are frequently used in industry:

(i) *Field erection and assembly drawings:* As their name signifies, both types of drawings are basically used to perform assembly operations.

(ii) *Machining Drawings and Schematic Diagrams:* Machining drawings are used to machine an item to its specified requirements. On the other hand, the schematic diagrams are used to show items' connections, characteristics and relationships.

(iii) *Detail Drawings:* The uses of these drawings are to provide necessary information to produce and inspect a product.

(iv) *Structural and Installation Drawings:* Structural drawings are sometimes also known as weldment drawings, which are highly specialized. These drawings are used to provide information to the manufacturing people so they can establish weld process specifications and standard times for manufacturing. Installation drawings are basically utilized to depict, with respect to other parts, where and how an item (or items) is mounted. In addition, these drawings show how the items in question are fixed with supporting structures.

(v) *Casting Drawings:* As their name suggests, these drawings are used for casting purposes. Foundries are the users of these drawings.

14.3   USERS OF DRAWINGS

There are various users of engineering drawings in industry, including:

(i) Management

(ii) Production and inventory control departments or groups

(iii) Marketing and purchasing departments or groups

(iv) Research and development, and design engineering departments or groups

  (v) Manufacturing and quality control departments or groups
 (vi) Publications and customers services departments or groups
(vii) Shipping and receiving departments or groups

14.4 USES OF ENGINEERING DRAWINGS

Engineering drawings have many uses, some of which are as follows:
   (i) Product maintenance
  (ii) Product modification
 (iii) Product design evaluation
  (iv) Research and development documentation
   (v) Identification of product components
  (vi) Product manufacuring, testing and installation
 (vii) Convincing management and customers
(viii) Reliability evaluation
  (ix) Component classification
   (x) Cataloging
  (xi) Advertisement literature preparation
 (xii) Maintenance manual preparation
(xiii) Product selling price setting
 (xiv) Product cost evaluation

14.5 DRAWING OFFICE

According to Reference [3], the principal objective of an engineering drawing office is two-fold:

   (i) To clothe technological ideas in hardware
  (ii) To prepare understandable instructions to others so that they are able
        to assemble and install engineering equipment, instruments, etc.

In order to fulfill such objectives, the drawing office has to carry out various
functions. Some of them are as follows:

   (i) Formulating objectives and policies
  (ii) Planning work
 (iii) Scheduling work
  (iv) Control

   (v) Supervision
  (vi) Personnel selection
 (vii) Drafting
(viii) Liaison with others
  (ix) Releasing drawings

14.5.1   Drawing Office Liaison with Other Departments

In order to function effectively, it is essential for a drawing office to have close relationships with the various groups or departments. Better communication or contacts with others will always be helpful in producing better quality end results or products. Therefore, the drawing office generally has close links with the following departments or bodies [4]:

1. Test engineering
2. Production engineering
3. Customers
4. Marketing
5. Plant engineering
6. Development engineering
7. Regulations

14.5.2   Drawing Office Supervision

Effective supervision of the drawing office is necessary to produce correct drawings on specified time at minimum cost. According to Reference [5], virtually every work situation suffers from the following principal factors:

  (i) Poorly clarified boundaries of employee job roles
 (ii) Poor understanding of actual expectation from employees
(iii) Subjectively evaluated performance

It is understood that employees perform best when their tasks are properly defined, and when they know what management expects from them, what the measurements of performance are, and what the end rewards are for a job well done.

Therefore, to achieve effective drawing office performance, consideration must be given to the following factors:

  (i) Concise and clear work assignments
 (ii) Assignments record keeping and monitoring
(iii) Clarification of supervisory authority

(iv) Feedback regarding the assignments

(v) Weekly work assignment progress monitoring in the presence of other group members

14.5.3 Common Drawing Office Efficiency Problems

Just like any other department, the drawing office is also prone to efficiency problems. The drawing office efficiency problems can be accredited to various factors. Some of them are as follows:

(i) Lack of product standards

(ii) Planning of drawing office

(iii) Drawing office "down time"

(iv) Scheduling

(v) Changes in design during working drawings

14.5.4 Omissions and Frequently-Occurring Errors in Working Drawings

A drawing office produces various types of new drawings. In working drawings, various frequently-occurring errors are found. Regarding this aspect, a British construction research group conducted a survey in 1973. The objective of this survey was to find flaws in working drawings. The results of this study indicated that specific types of flaws were frequently occurring in the working drawing. The main results of these findings are categorized into four classes, as follows [5]:

(i) *Class I:* These errors are associated with notation. Both studies confirmed the frequent inconsistency in terminology and abbreviations in the same set of drawings, the too specific materials and construction notation in drawings, the incomplete notation, the inadequate reference notes, and the too small lettering used in drawings.

(ii) *Class II:* This category contains errors due to omissions. Various types of omissions in drawings are cited in Reference [5].

(iii) *Class III:* This class is concerned with technical errors. Various examples of such errors occurring in construction drawings are given in Reference [5].

(iv) *Class IV:* This class of errors is associated with dimensioning. According to the findings of the surveys, between 40 and 50% of individual drawings have errors in dimensioning, or empty spots. Examples of the oversights associated with construction drawings are briefly described in Reference [5].

14.5.5   Common-Sense Drafting Time-Saving Practices

The drawing office's function is to produce drawings. Therefore, management must be aware of the areas which will help to reduce drafting time. The following factors will help to save drafting time [6]:

    (i)   Draw only those dotted lines which are necessary.

  (ii)   Eliminate repetitive data by making use of notes.

 (iii)   Simplify dimensioning by using datum lines.

 (iv)   Eliminate unnecessary centre lines.

  (v)   Avoid duplication of dimensions.

 (vi)   For clarity, make use of cross-sectioning.

(vii)   Reduce or omit thread detail.

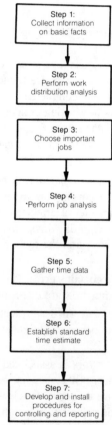

Figure 14.1.
Steps for implementing a work measurement program.

(viii) Eliminate detail which is repetitive and elaborate.

(ix) In the case of symmetrical objects, depict or show partial views only.

(x) If possible, avoid making large drawings.

(xi) Use hand or machine lettering when it is necessary.

14.5.6 Estimation of Drafting Time

Just like any other area, work measurement techniques can be utilized to obtain drafting time estimates. The subject of work study is described in Chapter 20 of this book. To develop time standards for drafting work, work measurement methods, motivational techniques and employee participation should be combined to obtain good performance results. According to Reference [7], this approach was already practiced at TRW with promising results. Once the drawing time standards are established they can be used as the basis for:

(i) Assigning drafting manpower duties; budgeting labor and cost.

(ii) Estimating, controlling and reducing cost.

(iii) Measuring drafting room or individual drafting person performance.

(iv) Forecasting the demand for drafting manpower.

(v) Improving methods and procedures of drafting.

(vi) Scheduling drawings.

A seven-step traditional procedure for implementing a work measurement program is shown in Figure 14.1. The steps of a work measurement program, which combines work measurement techniques, motivational techniques and employee participation, are given in Reference [7]. This program was developed at TRW.

14.6 SIMPLE RULES FOR THE DRAFTING MANAGER WHEN PRODUCING ORIGINAL DRAWINGS

These rules are very useful in producing original drawings [8]. Therefore, a drafting manager must take them into consideration on such occasion. These rules are as follows:

(i) Familiarize oneself with the company's methods for reproducing original drawings.

(ii) Familiarize oneself with the firm's reproduction equipment.

(iii) Make use of functional drafting methods.

(iv) In the beginning, with the aid of an original drawing, find out the fastest method of developing it.

(v) Avoid being over-functional to the point that the drawing becomes misleading.

(vi) If a drawing can be traced then avoid drawing it.

(vii) Before starting to develop a drawing through conventional means, make use of any possible time-saving reproduction technique.

14.7 WAYS TO REDUCE DRAFTING COSTS

Just like other departments and organizations, one of the drafting department's objectives is to reduce cost. There are various ways and means to cut down in drafting cost. According to an expert of in-drafting-room procedures, the following points are helpful in cutting costs [9]:

(i) *Clarify to the concerned drafting persons about the accuracy requirement in drawings:* Sometimes the accuracy of the drawings may not be that important, and therefore free hand sketches may be quite acceptable. In this situation the freehand sketches will be less costly to draw.

(ii) *Make sure that work environment is comfortable:* A good work environment will help to improve department efficiency. For example, fast-adjusting and easy-shift boards, and good chairs will be useful in cutting down costs.

(iii) Make sure that all commonly occurring materials are preprinted.

(iv) In the situation where one has to show changes in the existing equipment, make use of photographs instead of drawings, because the changes on the photoprint can be identified by marking.

(v) To reduce tracing time, make use of intermediates.

(vi) Make use of templets for a symbol with simple lines because they are more efficient than the tracing techniques or appliques.

(vii) Whenever it is possible, take advantage of new developments in paper used for drafting.

(viii) Take advantage of special or conventional typewriters instead of using lettering pens because typewriters are more efficient and cleaner than their counterparts, the lettering pens.

(ix) Whenever it is convenient, make use of drafting machines which fully perform the work of protractor, scale, triangles, etc.

(x) For frequently-occurring parts, prepare tracing templets.

(xi) For commonly-occurring symbols and diagrams, make use of self-adhesive appliques and rubber stamps.

(xii) Take advantage of newly-developed pens and pencils.

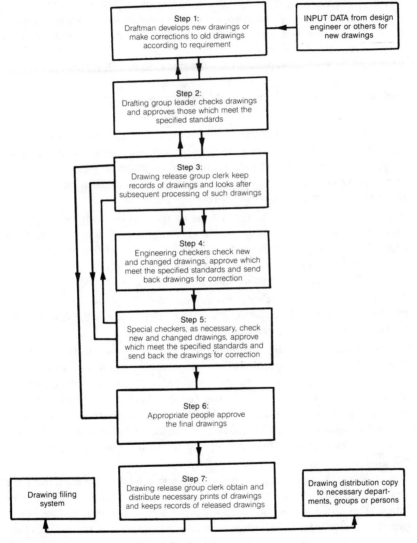

Figure 14.2.
Flow chart of a simplified approach for releasing engineering drawings.

14.8   AN APPROACH FOR RELEASING ENGINEERING DRAWINGS

Engineering drawings play an important role in developing new products. Therefore, a proper engineering drawing release procedure is needed to assure that the correct drawings are released for the product development. Furthermore, proper records are kept of drawings' whereabouts, so that the released drawings can easily be retrieved. In Figure 14.2, a flow chart of a simplified procedure for releasing engineering drawings is shown [10]. However, the release procedure may vary somewhat from company to company, due to the nature of the product and drawings. The basic steps of the procedure are briefly described as follows:

Step 1: In this step, the draftsman prepares a drawing with the aid of the specified data, or he (or she) carries out the necessary changes to the already prepared drawing as specified by the drafting group leader, engineer and special checkers. Once a drawing is fully completed, the draftsman passes on the drawing to his group leader for approval.

Step 2: The group leader or the supervisor receives the fully accomplished drawing and checks it, from the functional design aspect, for the correctness of necessary changes. However, if the supervisor is dissatisfied with the drawing, he passes it back to the draftsman for necessary action. When the supervisor is fully satisfied, he approves and forwards it to the release group clerk.

Step 3: In this step, the release group clerk looks after the drawing for the checking and approval by others. The release clerk records the drawing and forwards it to the engineering checkers.

Step 4: When the drawing is received by the engineering checkers, they examine it from various aspects, for example, exactness of dimensions, production design, conformance to drafting standards, correctness of changes made to the drawing, etc. However, if the engineering checkers are dissatisfied with the drawing, then the drawing is sent back and ultimately reaches the draftsman for changes or corrections. When the engineering checkers are fully satisfied with the drawing, they approve and return the drawing to the release clerk.

Step 5: When the release clerk receives the approved drawing back from the engineering checkers, he or she forwards the drawing to the special checkers group. This group may include stress analyst, weight engineer, tool engineer, production engineer, etc. However, the type and number of special checkers needed depends upon the nature of the product the drawing represents. All these special checkers examine the product from the aspect of their specialty. At this stage, if any change in the drawing is needed then the drawing is sent back and the draftsman makes necessary changes. When the special checkers are fully satisfied, they approve the drawing and return it to the release group clerk.

Step 6: Once the drawing, approved by the special checkers, is received by the release clerk then it is forwarded by the same clerk for the final engineering approval. This approval is usually obtained from the chief engineer or chief project engineer for new drawings. However, for the modified drawing, the final approval may be obtained from people such as the project engineer or chief draftsman.

Step 7: After the final engineering approval, the engineering release clerk obtains the necessary prints for the drawing. Once this is accomplished, the same clerk forwards the prints of the drawing to the necessary destinations and records their locations. In addition, the clerk sends one copy of the drawing to the master files.

14.9 MANAGEMENT OF DRAWING CHANGES

When producing new engineering products, changes to the original drawing may be required due to various reasons such as changes in customer requirements, to correct original drawings, etc. Thus, to handle such changes, a procedure has to be devised so that the required changes are made efficiently and accurately. Therefore this section briefly describes a procedure to handle such changes [11].

Step 1: The drafting supervisor assigns the task to a draftsman, who in turn contacts drawing release groups to gain more information on the requested change and any other outstanding changes to be incorporated. In addition, the draftsman investigates whether the changes in question will effect interchangeability. If so, special care is needed to handle those changes [11].

Step 2: In this step, all the required changes are made to the drawing by the draftsman. Furthermore, in the drawing title block the revision is recorded.

Step 3: The draftsman prepares the notice of change and forwards that notice, the drawing, the information on all changes made, and the back-up data to the group supervisor.

Step 4: The group supervisor checks the changes and approves if they are up to his or her satisfaction. Furthermore, he or she prepares a "release request" and forwards the changed drawings and the other necessary accompanying documents to the release group.

Step 5: The release group forwards the concerned documents to checkers and others for the checking approval. Afterward, the group clerk forwards the drawing and other documents to the draftsman for checker's corrections, if any. The draftsman incorporates the checker's corrections.

Step 6: The drawing release group makes prints of the corrected and approved drawing. Fnally, the group clerk files the copy of the drawing and distributes the remaining copies to the concerned destinations.

14.10   ENGINEERING DRAWING CHECK LIST

This section presents some of the important points which are to be checked for correctness when applicable in engineering drawings. These are as follows:

   (i) *Dimensions:* This is to find out if the necessary dimensions are shown.

  (ii) *Accessibility:* This is concerned with investigating if there is enough accessibility for riveting and welding.

 (iii) *Scale:* This is concerned with finding out if the drawing is drawn to scale.

 (iv) *Production Facilitation:* The objective of this checkpoint is to find out if a part is designed to take care of production.

  (v) *Finishing Allowances:* This is concerned with examining if the necessary finishing allowances are taken care of.

 (vi) *Standard Parts Specification:* The objective of this point is to see if the standard parts are properly specified.

 (vii) *Usage of Standard Parts:* This is to see if the standard components are used whenever it is possible.

(viii) *Drawing General Appearance:* Here, one examines if the drawing's general appearance is satisfactory.

 (ix) *Drafting Standards:* This is concerned with finding out if the drawing is prepared according to company or other drafting standards.

  (x) *Cleanliness of Tracing.*

 (xi) *Legibility and Satisfactory Reproduction of Lines, Figures and Letters.*

 (xii) *Completion of the Title Block:* Here one examines whether the title block is correctly filled and complete.

(xiii) *Provision for Notes, Charts and Data.*

(xiv) *Comparison with Similar Drawings Produced Earlier:* This point is concerned with examining how this drawing compares with similar drawings produced earlier.

14.11   SUMMARY

This chapter briefly discusses the various aspects of engineering drawings, with emphasis on managerial aspects. The chapter begins with briefly describing the historical aspect of the drawings and goes on to discuss the types of technical illustrations and drawings. Furthermore, the fourteen different applications of engineering drawings are given.

The various aspects of a drawing office and related areas are covered in the chapter. For example, functions of a drawing office, drawing office liaison with other departments, drawing office supervision and efficiency problems. In addition, the commonly occurring errors in working drawings, drafting time-saving practices, and estimation of drafting time are also discussed.

The chapter briefly discusses the various ways to cut down drafting costs. Twelve ways to cut down cost are presented.

A seven-step approach for releasing engineering drawings is described. In addition, a procedure for handling engineering drawings changes is discussed. In the final part of the chapter, an engineering drawing checklist is presented.

Important articles related to the material presented in the chapter are listed in the reference section.

14.12 EXERCISES

1. What are the application areas of engineering drawings?
2. What are the main types of engineering drawings?
3. Discuss the functions of a drafting department.
4. Describe briefly the following:
   (i) Installation drawings.
   (ii) Schematic diagrams.
   (iii) Four different users of drawings.
   (iv) The principal objective of an engineering drawing.
5. Describe the uses of engineering drawings in the production and marketing departments.
6. What are the obstacles encountered in drawing office efficiency?
7. What are the errros which commonly occur in working drawings?
8. What are the drafting practices which help to cut down the drafting cost.
9. Describe a procedure for handling engineering drawing changes.

14.13 REFERENCES

1. Giesecke, F. E., A. Mitchell, H. C. Spencer, I. L. Hill, R. O. Loving and J. T. Dygdon. *Engineering Graphics*. New York:Macmillan Publishing Co., Inc. (1981).
2. Rowbotham, G. E., ed.-in-chief. *Engineering and Industrial Graphics Handbook*. New York:McGraw-Hill Book Company (1982).
3. Gregory, S. A. *The Design Method*. London:Butterworths (1966).
4. Matousek, R. *Engineering Design*. New York:John Wiley & Sons (1963).
5. Stitt, F. A. *Systems Drafting*. New York:McGraw-Hill Book Company (1980).
6. Karger, D. W. and R. G. Murdick. *Design Drafting Function in Management Guide for Engineers and Technical Administrators*. N. P. Chrionis, ed. New York:McGraw-Hill Book Company (1969).

7. Beck, G. R. "How to Get Realistic Drafting–Time Estimates," in *Management Guide for Engineers and Technical Administrators*. N. P. Chironis, ed. New York:McGraw-Hill Book Company (1969).

8. Irwin, W. J. "More Hints for the Drafting Manager," in *Management Guide for Engineers and Technical Administrators*. N. P. Chironis, ed. New York:McGraw-Hill Book Company (1969).

9. Gagne, A. F. "Drafting Costs," in *Management Guide for Engineers and Technical Administrators*. N. P. Chironis, ed. New York:McGraw-Hill Book Company (1969).

10. Thompson, J. E. "Release and Control Procedures for Engineering Drawings," in *Management Guide for Engineers and Technical Administrators*. N. P. Chironis, ed. New York:McGraw-Hill Book Company (1969).

11. Thompson, J. E. "Handling Drawing Changes in Small Companies," in *Management Guide for Engineers and Technical Administrators*. N. P. Chironis, ed. New York:McGraw-Hill Book Company (1969).

CHAPTER 15

Value Engineering and Configuration Management

15.1 INTRODUCTION

The topics of value engineering and configuration management are of considerable interest to engineers because they may get involved sometime during their careers directly or indirectly with both these areas. Furthermore, the importance of these two subjects has been following an increasing trend in recent times. This may be due to tight economic environment, competition, expensive and complex systems, etc.

In the case of value engineering, it may be said that it is principally a function-oriented discipline. Furthermore, its principles pertain to cost prevention and reduction. At present, several different definitions are used to describe value engineering; however, the central theme of each of these definitions is that value engineering has two objectives: reducing cost and improving quality by the means of better design and manufacturing techniques [1]. The history of value engineering goes back to 1947 when it was first applied at General Electric, under the title of value analysis, by L.D. Miles. The U.S. Navy enhanced its importance by establishing several value engineering departments in the fifties. According to Reference [2], in the United States Department of Defense alone, the audited savings of more than 1.1 billion dollars was calculated from the fiscal years 1963 through 1966. Two other important events in the history of value engineering are as follows:

(i) The American Society of Value Engineers was established in 1959.

(ii) *The Journal of Value Engineering* was started in 1962.

Today, many universities and colleges teach courses on value engineering.

In the case of configuration management, it may be simply stated that it is the ways or means by which engineering equipment and documentation are kept mutually identified [3]. The history of modern configuration management goes back to the 1950s, when the people involved in the missile launch program became aware of its need because they felt there was insufficient documentation associated with the changes made to the product. However, a formal document on the subject entitled "Configuration Management During the Development and Acquisition Phases," AFSCM 375-1, was published in 1962

271

by the United States Air Force. Since then, several military and other publications on the subject have appeared.

In this chapter, the topics of value engineering and configuration management are discussed in detail.

15.2 VALUE ENGINEERING

This is a cost-quality inclined discipline which is concerned with questions, evaluation and analysis in order to enhance the quality of product and increase the ultimate profits of the organization. According to Reference [2], the following are the reasons given by organizations for not having the value engineering program:

  (i) Our organization performs service-oriented tasks, whereas value engineering is concerned with hardware.

  (ii) There is a large variation in our product price, quality, size, use, etc.

  (iii) We do not practice value engineering because the size of our organization is too small.

  (iv) Our organization is basically concerned with research and development, and we are producing highly sophisticated products for the first time. Therefore, the size of the production quantity will not allow us to realize in full the value engineering profits.

  (v) Our product uses a large proportion of components which are purchased from others. Therefore, the value engineering practice will not be a valuable venture.

Various aspects of value engineering are discussed in the following sections.

15.2.1 Poor Value Factors

Here one may say that the prime reason for poor value is the non-presence of systematic effort to achieve high value. To obtain high value, the counter-offensive must be organized in such a way so that the following factors are eliminated:

  (i) Lack of new ideas

  (ii) Non-availability of proper information

  (iii) Decisions made on a temporary basis

  (iv) Weakness in seeking advice from others

  (v) Tendency to resist change

  (vi) Personal inertia

15.2.2   Tasks of a Value Engineer

For a specific task, a value engineer performs the following steps:

(i) Selects the most suitable materials for the task or job in question

(ii) Chooses the most appropriate processes of manufacturing

(iii) Chooses the most suitable equipment and tools to fabricate the product in question

(iv) Chooses the most appropriate design approaches

(v) Identifies those cost factors which are the causes for the product value reduction

(vi) Identifies those quality factors which play an important role in reducing the value of the product

(vii) Replaces those items and procedures which reduce the product value with those which increase the value of the product. For example, inadequate specifications, low quality design and poor approaches may be the causes which reduce product value.

The value engineer should ask the questions given in Figure 15.1 in order to fulfill the above objectives. These questions are concerned with cost, identification, function, alternatives, quality improvement through alternatives, and cost of alternatives. To measure the performance of the product in the

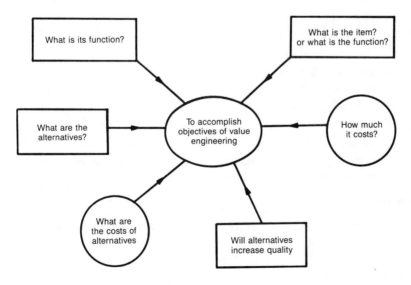

Figure 15.1.
Questions for fulfilling value engineering objectives.

market or in the field, the value engineer should look into the following:

 (i) Responses or reactions of the customers or buyers
 (ii) Product's position in the market due to competition (What are the causes for its weaknesses?)
 (iii) Comparative analysis
 (iv) Sales potential of the product
 (v) High profit margin products

15.2.3   Phases of Value Engineering

This section briefly discusses the phases of performing value engineering study. According to Reference [2], the study consists of six phases which are concerned with information (Phase I), functional analysis (Phase II), speculation (Phase III), evaluation (Phase IV), program planning and presentation (Phase V), and implementation (Phase VI). All of these phases are described below.

PHASE I

This phase is concerned with collecting the essential information on the following items:

 (i) Objective
 (ii) Use
 (iii) Physical needs
 (iv) Environmental needs
 (v) Problem areas of concern
 (vi) Economic aspects such as useful life period, estimated manufacturing cost incurred presently, delivery date of the first production unit, etc.
 (vii) Support needs
 (viii) Testing needs
 (ix) Liaison persons in departments such as design and development, production, purchasing, etc.

PHASE II

This phase deals with functional analysis. Two important questions asked during this phase are:

 (i) What is its function?
 (ii) What is the worth of that function?

To find solutions to the above questions, one has to identify function, estimate the worth of that function, and obtain data for the actual cost of the function.

PHASE III

This phase is known as the speculation or creative phase and is basically concerned with creativeness. In this phase, one seeks the answer to questions such as "What are the other two ways to perform the same function?" The two mental processes involved in this phase are the creative and the judicial.

PHASE IV

This phase is known as the evaluation phase. Basically, in this phase one is involved in developing choices among several alternatives. To accomplish this, the following two steps are performed:

  (i) Feasibility analysis
  (ii) Resource analysis

PHASE V

This is known as the program planning and presentation phase. In this phase, the value engineer plays the role of a salesman who has to sell value engineering change proposals to others.

PHASE VI

This is the last phase of the value engineering study and is known as the implementation phase. This is concerned with the implementation of value engineering change proposals. Before these proposals are implemented, the value engineer must get the written approval of the management in order to be assured of its backing. Otherwise, value engineering proposals implementation may run into problems. Once these proposals are fully implemented, one must follow up to monitor the results of the value engineering study.

15.3 CONFIGURATION MANAGEMENT

Prior to World War II, engineering products were less sophisticated and the changes to those products were made less frequently. However, in the post-World War II era, engineering products have become more sophisticated and the frequency of changes made to products is on the increase. These changes are mainly concerned with size, performance, design, new parts, software, appearance, etc. Configuration management is the discipline concerned with these changes. Therefore, this section presents various aspects of configuration management.

15.3.1   Configuration Management Achievements

The following are the principal reasons for configuration management implementation [4,5]:

(i) To realize total economy in money

(ii) To enhance efficiency in the conduct of the project

(iii) To realize total economy in time

By accomplishing the following through configuration management, the above objectives are automatically fulfilled:

(i) Selecting areas needing research to meet the goal of the project

(ii) Identifying the project principal goals

(iii) Imposing restrictions so that only the vital changes and revisions are made

(iv) Making sure that the product support in the design, testing and field operation phases will be implemented systematically with the availability of necessary manpower and components

(v) Formulating a program in which the resources within reach are identified, scheduled, etc.

(vi) Keeping track of those identification disciplines which are necessary for realizing the creation of documentation concerning drawings, data, etc., for describing fully the product and associated effort

15.3.2   Problem Areas for Configuration Management Application

This section presents difficulties which may exist in a company and can be overcome with the application of the configuration management principles. These difficulties arise in four areas: research, design and development engineering, and manufacturing.

In research, poor monitoring and control of changes, and poor direction from management are the main causes for the need of configuration management principles.

The following difficulties, which occur in design and development engineering, need the application of configuration management principles:

(i) Poor standardization program planning and documentation

(ii) Poorly monitored and controlled product design changes

(iii) Poorly defined product design and development objectives and procedures

(iv) Poor leadership of management

(v) Poor data retrieval and cataloguing system

(vi) Lack of communication with other concerned bodies

In the manufacturing area, configuration management can be applied to correct the following difficulties:

(i) Inefficiency in data storage and retrieval system

(ii) Control of revisions inadequacy

(iii) Inadequate compatibility between the engineering and design specification documentation

(iv) Inaccuracy in maintenance and spare procedures description data

The principal difficulties in the field support area are poor communication between users and equipment developers, poor equipment field modification documentation system, and inadequate control of product modification decisions.

15.3.3 Configuration Management Plan

This is composed of four phases known as project conceptual, definition, procurement, and field operation. The subcomponents of the procurement phase are the design and development stage, and production stage. The four phases end at characteristics, functional, product (the sub-baseline) and operational baselines. In other words, the conceptual, definition and procurement (acquisition) [design and development stage and product stage] phases terminate at the characteristics, functional and operational baselines, respectively. Because the procurement (acquisition) phase is divided into two portions, the design and development stage terminates at the product sub-baseline and the production stage at the operational baseline. The configuration management plan phases and their corresponding termination baselines are given in Table 15.1.

One should note here that any of the succeeding phases cannot start until that time when all the objectives denoted by the baseline in question are fully fulfilled.

Table 15.1 Configuration Management Plan Phases and their Corresponding Baselines or Points.

Configuration Management Plan Phase	Corresponding Termination Baseline or Point
a) Conceptual	a) Characteristics
b) Definition	b) Functional

c) Procurement
   (i) Design and Development stage
   (ii) Production stage
	c) Operational
	   (i) Product sub-baseline
	   (ii) Operational baseline

d) Field operation	d) Product scrapped

According to the degree of need, the disciplines associated with configuration management, such as identification, control and accounting, are applied within the framework of each of the above phases. These three are briefly described in the following sections [5].

IDENTIFICATION

The principal concern of identification in the concept formulation phase is that of the techniques or approaches, ideas, etc. Furthermore, the definition of the product in question is established in conceptual terms, and the objectives are thoroughly reviewed. More effort is spent on identification requirements in the conceptual phase than on control and accounting functions, or on identification in other phases.

During the definition phase, the identification implemented is more solid. Basically it translates the ideas of the previous phase into product specification documents, and other reports. The effort spent for identification during this phase is somewhat less than that spent in the concept formulation phase.

During the procurement phase, once the design and development stage is over, the effort for identification decreases rapidly.

The effort for identification is the lowest in the final phase (field operation phase), in comparison to all other phases. During this phase the basic identification effort is directed toward product modifications and other similar activities.

CONTROL

During the conceptual phase, control functions play a low-key role and basically are exercised to meet baseline objectives by directing concept formulation and research effort. In the remaining phases as concrete progress is made on the project, as expected, the effort on control activities increases. Generally, the more significant increase in effort related to control activities is experienced during the procurement phase.

ACCOUNTING

In order to provide effective logistic support, the accounting discipline is concerned with establishing records. More clearly, in simple terms, it is concerned with formulating an effective approach so that the product descriptions and changes are properly documented. Although accounting may be needed on sophisticated and larger programs in the definition phase, its principal usage is in the procurement and field operation phases.

Some of the advantages of the configuration management application are as follows:

(i) Accurate data retrieval facilitation

(ii) Effectiveness in resource channeling

(iii) Establishment of objectives for each of the four phases

(iv) Getting rid of redundancy in effort

For a detailed description of the above four benefits, if desired, consult Reference [4]. The discipline of configuration management is described in detail in Reference [6].

15.4 SUMMARY

This chapter briefly describes the disciplines of value engineering and configuration management. The histories of both subjects are briefly discussed. The chapter is divided into two parts, i.e., value engineering and configuration management.

In the value engineering portion, reasons for not having a value engineering program, poor value factors, tasks of a value engineer, and the six phases of value engineering are discussed.

In the second half of the chapter, configuration management objectives, accomplishments, problem areas, plan and techniques are described briefly.

References relating to each of these two disciplines are listed at the end of the chapter.

15.5 EXERCISES

1. Discuss the histories of value engineering and configuration management.

2. Write an essay on value engineering and configuration management principles.

3. What are the factors a value engineer should look into to evaluate performance of a product in the market?

4. What are the difficulties that may exist in a company and can be overcome with the application of configuration management?

5. Describe what is meant by the term "baseline"? Discuss at least two baselines.

6. Describe the procurement (acquisition) phase of the configuration management plan and the degree of application of identification control and accounting in that phase.

15.6 REFERENCES

1. Hayes, G. E. and H. G. Romig. *Modern Quality Control.* Encino, CA:Bruce: A Division of Benziger & Glencoe, Inc. (1977).

2. "Value Engineering," Engineering Design Handbook, U.S. Army Material Command, AMCP 706-104, Available from the National Technical Information Service, Springfield, VA (July 1971).

3. Juran, J. M., F. M. Gryna and R. S. Bingham. *Quality Control Handbook*. New York: McGraw-Hill Book Company (1979).

4. Hajek, V. G. *Management of Engineering Project*. New York:McGraw-Hill Book Company (1977).

5. Hanjek, V. G. *Configuration Management*. 30 Fleet Street, London, E.C.4:Industrial and Commercial Techniques, Ltd.

6. Samaras, T. T. and F. L. Czerwinski. *Fundamentals of Configuration Management*. New York:John Wiley & Sons (1971).

CHAPTER 16

Management of Product Assurance Sciences

16.1 INTRODUCTION

To produce reliable and sophisticated products at a minimum cost, the emphasis on product assurance sciences has been showing an increasing trend. According to Reference [1], about 10 to 15 percent of the product development money in some aerospace companies is spent on product assurance. Areas such as reliability, maintainability, quality control, and safety are the components of product assurance sciences.

Reliability is the probability that a product will perform its intended function satisfactorily for the specified period when used under the designed conditions [2]. The modern history of the reliability field begins with World War II and is associated with the development of V-1 and V-2 rockets. Here the reliability concept was applied by the Germans to improve reliability of their rockets. Ever since then, the field of reliability has matured to a level where it is either beginning to be or already has been identified into various subbranches. Most of the specialized branches of reliability are presented in References [2] and [3].

Maintainability is another area of product assurance sciences which is defined as the probability that a failed product will be restored to its satisfactory operational state within a specified downtime interval, when the maintenance is performed according to desired procedures and resources. In comparison to reliability, the history of modern maintainability is relatively new and goes back to the early fifties [4]. However, a maintainablity document entitled MIL-M-26512 was published by the United States Air Force in 1959. Ever since then, several authors and researchers have contributed to the field of maintainability. A list of references on the subject may be found in Reference [2].

Quality control, the third important area of assurance sciences, is a management function which is exercised by controlling the raw materials quality and manufactured items to prevent defective units production. The history of quality control goes back further than that of reliability and maintainability. In 1916, C. N. Frazee of Telephone Laboratories applied the statistical techniques for the first time to inspection problems. Another important event in the history of quality control took place in 1924 when W. A. Shewhart developed the quality control charts. Since then, a vast amount of literature on the subject has been published and quality control techniques are being applied across many diverse areas [5].

Safety, the last area of assurance sciences discussed in this chapter, is defined as independence from those conditions that can cause injury or death to human beings or damage to or loss of item [6]. Since the early fifties system safety has been receiving considerable attention from the United States Department of Defense. However it was not until 1962 that the first document entitled "System Safety Engineering for the Development of United States Air Force (USAF) Ballistic Missles" was published. MIL-STD-38130 was published by the USAF a year later in 1963. Since then several publications have appeared on the subject of system safety. A list of publications on the subject is given in Reference [2].

16.2   INTRODUCTION TO RELIABILITY AND RELIABILITY MANAGEMENT

This section briefly describes the basic reliability measures and networks and the various aspects of reliability management.

16.2.1   Reliability Measures

This section presents three formulas for calculating item reliability, mean time to failure, and failure rate (hazard rate). Derivations of these formulas may be found in Reference [3]. These formulas are as follows:

RELIABILITY

Reliability, $R(t)$, at time t of an item, is given by

$$R(t) = 1 - \int_0^t f(t)\, dt \qquad (16.1)$$

or

$$R(t) = e^{-\int_0^t \lambda(t)\, dt} \qquad (16.2)$$

where

$f(t)$ is the failure probability density function.
$\lambda(t)$ is the hazard rate or the time dependent failure rate.

To calculate an item's reliability, either one of the above equations can be utilized.

EXAMPLE 16.1

An electronic component's failure times are described by the following failure density function:

$$f(t) = \lambda\, e^{-\lambda t} \tag{16.3}$$

where λ is the constant failure rate of the component.

Calculate the component reliability for a 150 hour mission when the value of $\lambda = 0.005$ failure/hour.

Substituting Equation (16.3) into Equation (16.1) and integrating leads to

$$R(t) = 1 - \int_0^t \lambda\, e^{-\lambda t}\, dt$$
$$= e^{-\lambda t} \tag{16.4}$$

Utilizing the given data for λ and t in Equation (16.4) results in

$$R(150) = e^{-(0.005)\,(150)}$$

$$= 0.4724$$

Thus the reliability of the component is 0.4724.

EXAMPLE 16.2

An electric motor hazard rate is

$$\lambda(t) = \lambda \tag{16.5}$$

Calculate the motor reliability for a 100 hour mission. Assume the value of the motor constant failure rate $\lambda = 0.0002$ failure/hour.

Thus substituting Equation (16.5) into Equation (16.2) and integrating leads to

$$R(t) = e^{-\int_0^t \lambda\, dt} = e^{-\lambda t} \tag{16.6}$$

Equation (16.6) is the same as Equation (16.4). Thus, making use of the specified data in Equation (16.6) yields

$$R(100) = e^{-(0.0002)(100)}$$

$$= 0.9802$$

The electric motor reliability is 0.9802.

MEAN TIME TO FAILURE (MTTF)

This is defined by

$$MTTF = \int_{0}^{\infty} R(t)\, dt \tag{16.7}$$

EXAMPLE 16.3

Reliability, $R(t)$, of a mechanical component is defined by

$$R(t) = e^{-\lambda t^{b}}; \text{ for } \lambda > 0, b > 0, t \geq 0 \tag{16.8}$$

where

λ is the reciprocal of the scale parameter.
b is the shape parameter.
t is time.

Obtain an expression for the component mean time to failure. If $b = 1$ and $\lambda = 0.005$ failure/hour, then calculate the component mean time to failure.

Incorporating Equation (16.8) into Equation (16.7) and integrating yields:

$$MTTF = \int_{0}^{\infty} e^{-\lambda t^{b}}\, dt$$

$$= \frac{\Gamma(1/b)}{b(\lambda)^{1/b}} \tag{16.9}$$

where

$$\Gamma(k) = (k - 1)!, \text{ for positive integer values} \tag{16.10}$$

and

$$\Gamma(\tfrac{1}{2}) = \sqrt{\pi} \tag{16.11}$$

Using the specified values of λ and b in Equation (16.9) yields

$$MTTF = \frac{\Gamma(1)}{(.005)} = \frac{1}{0.005} = 200 \text{ hours}$$

The component mean time to failure is 200 hours.

HAZARD RATE

This is defined by

$$\lambda(t) = \frac{f(t)}{R(t)} \tag{16.12}$$

where

$\lambda(t)$ is the item hazard rate.
$f(t)$ is the item failure density function.
$R(t)$ is the item reliability function.

EXAMPLE 16.4

An electric power generator's failure times are represented by the following probability density function.

$$f(t) = \lambda\, e^{-\lambda t} \tag{16.13}$$

where t is time and λ is a parameter.
Obtain an expression for the generator failure rate.
Equation (16.13) is the same as Equation (16.3). Thus from Equation (16.4), the generator reliability is given by

$$R(t) = e^{-\lambda t} \tag{16.14}$$

Incorporating Equations (16.13) through (16.14) into Equation (16.12) yields

$$\lambda(t) = \frac{\lambda\, e^{-\lambda t}}{e^{-\lambda t}} = \lambda \tag{16.15}$$

Thus, λ is the generator failure rate.

16.2.2   Reliability Networks

This section is concerned with the reliability evaluation of two basic networks frequently encountered in the reliability field. These are presented in the following sections.

SERIES NETWORK

This is the simplest configuration used to evaluate the reliability of engineering systems. In this configuration it is assumed that the network (system) is composed of N independent units in a series. If any one of the units fails, the system (network) fails. In other words, all the network units have to function successfully for system success. The series system block diagram is shown in Figure 16.1, according to which reliability, R_s, is given by

$$R_s = R_1 \cdot R_2 \cdot R_3 \cdot --- \cdot R_N \qquad (16.16)$$

where R_i is the ith unit constant reliability (probability of success); for $i = 1,2,3, ---, N$.

EXAMPLE 16.5

An aircraft has two independent and identical engines. Both engines are needed for the aircraft to fly successfully. Each engine's probability of success is 0.95. In other words, it will function normally 95 times out of 100. Calculate the reliability of the aircraft flying successfully.

Since the aircraft has two identical engines, for Equation (16.16) we have the following data:

$$R_1 = R_2 = 0.95 \text{ and } N = 2$$

Thus substituting the above data into Equation (16.16) results in

$$R_s = R_1 \cdot R_2 = (0.95)(0.95) = 0.9025$$

The probability of the aircraft flying successfully is 0.9025 or 90.25 percent.

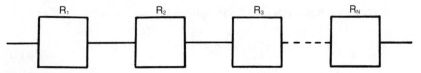

Figure 16.1.
Series network block diagram.

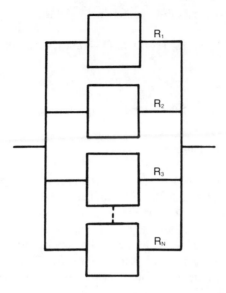

Figure 16.2.
Parallel network block diagram.

PARALLEL NETWORK

This network (system) is assumed to consist of N active and independent units in parallel as shown in Figure 16.2. At least one unit must operate normally for system (network) success. The system fails only when all its units fail. This configuration is frequently used to improve system reliability where a single unit's reliability cannot meet the specified reliability requirement.

Figure 16.2 network reliability, R_p, is given by

$$R_p = 1 - (1 - R_1)(1 - R_2)(1 - R_3) --- (1 - R_N) \qquad (16.17)$$

where R_i is the ith unit reliability (the probability of operating normally); for $i = 1,2,3, ---, N$.

EXAMPLE 16.6

An aircraft has three independent and identical engines. At least one engine must function normally for the aircraft to fly successfully. Each engine's probability of success is 0.9. In other words, in 90 times out of 100 the engine will operate normally. Calculate the reliability of the aircraft flying successfully.

In this example, the data is specified for the following items of Equation (6.17):

$$R_1 = R_2 = R_3 = 0.90 \text{ and } N = 3$$

Utilizing the above data in Equation (6.17) yields:

$$R_p = 1 - (1 - 0.90)(1 - 0.90)(1 - 0.90)$$

$$= 0.999$$

Thus the probability of the aircraft flying successfully is 99.9 percent.

16.2.3   Reliability Management

Reliability is considered an important parameter when developing new engineering systems, especially for use in national security, aerospace, etc. Generally, the reliability requirements of such systems are specified in the design specification. Therefore, to meet the specified reliability of a product, the reliability associated activities have to be managed in such a manner so that the end targets are satisfied at minimum cost. Therefore, this section is concerned with the reliability management associated aspects. Some of these aspects are described in the following sections.

RESPONSIBILITIES OF THE RELIABILITY ENGINEERING DEPARTMENT

The main objective of any reliability engineering department is to assure that the company manufactures a reliability product at the lowest cost. To accomplish this objective the department has the following main responsibilities:

   (i) Reliability planning and allocation
  (ii) Reviewing product specification and design from the reliability standpoint
 (iii) Providing reliability input to product design specification and proposal
 (iv) Reliability evaluation
  (v) Failure data collection and analysis
 (vi) Reliability growth monitoring
(vii) Failure data reporting
(viii) Reliability demonstration
 (ix) Reliability research and training
  (x) Reliability activity auditing (i.e., own company, sub-contractor, vendor, etc.)
 (xi) Reliability consulting to other groups, departments and organizations
(xii) Preparing reliability manual, documentation, budgeting and warranties

These responsibilities may vary from company to company depending on the size of the organization, nature of product, thinking philosophy of management, etc. Some of these selective responsibilities are described in the following sections.

RELIABILITY MANUAL

Manuals are commonly used in engineering organizations to outline organizational structure and authority, define functional relationships between departments, groups and divisions, define responsibilities, outline work procedures, etc. Therefore, a reliability engineering department generally has its own manual. The following items should be incorporated into the reliability manual:

(i) Reliability policy, organizational structure and responsibilities

(ii) Reliability evaluation techniques, procedures, models, etc.

(iii) Reliability procedures concerning component selection, design review, data collection and analysis

(iv) Reliability testing and demonstration procedures and methods

Because of the variation in reliability responsibilities, the reliability manual may vary from company to company.

REQUIREMENTS IN DESIGN SPECIFICATION

Ensuring that reliability requirements are properly specified in the product design specification is one of the responsibilities of the reliability engineering department. A typical reliability department of an aerospace company may have the following four responsibilities to satisfy reliablity requirements in design specification:

(i) Make certain that the reliability needs are incorporated into the system design specification as well as into the sub-contractor's portion of the specification

(ii) Make sure that the customer's reliability requirements are fully satisfied during the system design and development phase

(iii) Make certain that the reliability measures committed in the technical proposal can be fulfilled satisfactorily by the organization

RELIABILITY BUDGETING

Reliability budgeting is one of the vital responsibilities of reliability management.

Preparation and getting approval of a budget are important tasks which confront every reliability manager. The budget requirement has to be forecasted in such a way that it gets the approval of the top management and fully satisfies the need of the forthcoming period. Furthermore, in the eyes of the customer, the percentage of research and development money allocated towards reliability indicates the importance the company gives to reliability. The U.S. government missile contractors' survey reported in Reference [7] indicates that the companies participating in the survey spent between 0.3 percent and 10 percent of their research and development funds on reliability.

16.3   MAINTAINABILITY MANAGEMENT

Just like reliability management, maintainability management is concerned with accomplishing various similar tasks. Basically, these tasks fall into five categories [4] which are administration, coordination, design, analysis, and documentation. These five are discussed separately in the following sections.

16.3.1   Administration

This is concerned with tasks such as organizing, staffing, assigning responsibilities, preparing budgets, schedules and maintainability program plans, participating in design reviews, and developing and executing procedures and policies.

16.3.2   Coordination and Documentation

Both these functions play an important role in the management of maintainability. The coordinating aspect deals with interfacing with customers, sub-contractors, system engineering, etc. On the other hand, the documentation is concerned with documenting the results of design reviews, maintainability data, other information related to maintainability, etc.

16.3.3   Design

This is another important aspect of maintainability management. Here one looks into equipment design from different maintainability perspectives so that the logistic need is reduced to a minimum. In the design function, the main activities of concern are participating in the design reviews, preparing design reports related to maintainability, monitoring equipment design, preparing maintainability documents for internal use, approving design drawings, and providing consultation to others.

16.3.4   Analysis

This is composed of tasks such as maintainability allocation and prediction, maintainability trade-off analysis, maintainability demonstration, reviewing documents such as specifications, proposals, etc., from the maintainability aspect, and participating in system and engineering maintenance analysis from the maintainability standpoint.

16.4   QUALITY CONTROL MANAGEMENT

Quality control is one of the important elements of the product assurance sciences. Following are most of the tasks generally assigned to the quality control department [5]:

   (i)   Perform quality cost analysis
   (ii)   Plan for quality and inspection
   (iii)   Develop motivational programs
   (iv)   Develop quality objectives and policies
   (v)   Inspection
   (vi)   Monitor vendors, sub-contractors, etc.
   (vii)   Participate in design reviews
   (viii)   Analyze field complaints
   (ix)   Process control
   (x)   Analyse defect causes
   (xi)   Develop defect prevention programs
   (xii)   Statistical methodology
   (xiii)   Evaluate measuring instruments and equipment
   (xiv)   Budget for quality
   (xv)   Training
   (xvi)   Prepare quality control manual

16.4.1   Quality Manager's Role

A quality control manager plays various important roles in producing good quality products at minimum cost. According to Reference [5], the quality control manager basically plays the following roles:

   (i)   Planning
   (ii)   Coordination
   (iii)   Analytical
   (iv)   Consulting
   (v)   Liaison
   (vi)   Inspection

All the above roles are self-explanatory; however, a brief description on all of them is given in Reference [5].

16.4.2   Manual for Quality Control

This is an important document, and all quality conscious companies usually have such documents which contain the information on company policies and procedures concerning the quality of the product. The manual is a useful tool for [8]:

(i) training quality personnel and others because it can be utilized as a textbook.

(ii) providing reference for the policies and procedures.

(iii) auditing current quality practices because it provides a reference base.

(vi) providing a precedent to make decisions in the future.

(v) helping the continuity of operations despite the shortcomings of manpower turnover.

A quality control manual usually contains general information and managerial procedures information. The general information is associated with the definitions, responsibilities, organization charts and quality policies. The managerial procedures are concerned with quality costs, marketing, vendor quality control, inspection, testing, defect prevention, new-product introduction, statistical methodology, personnel, measuring equipment, etc.

16.5 SYSTEM SAFETY MANAGEMENT

System safety is another important area of product assurance sciences and is considered when developing and producing new engineering equipment. Some of the functions of system safety are as follows:

(i) Establishing the accident/safety related data bank

(ii) Taking part in product design reviews

(iii) Establishing system safety management objectives, plans and product design requirements

(iv) Performing safety analysis

(v) Determining emergency procedures

(vi) Preparing procedures for accident investigation and participating in the investigation

(vii) Training

(viii) Liaison with outside bodies

Tasks concerning system safety are integrated into the product during the product life cycle which is composed of conceptual, design and development, manufacturing, and operational phases.

16.5.1 System Safety Analysis Procedures

This section briefly presents two approaches for performing system safety analysis. These are as follows [5].

APPROACH I

This is a systematic procedure composed of five steps which becomes useful when designing a new product. These steps are shown in Figure 16.3.

APPROACH II

This is concerned with determining the occurrence probability of a hazardous event. Thus, the probability P_{use}, of an unsafe event occurrence is given by

$$P_{use} = \alpha TF; \text{ for } P_{use} \leq 1 \qquad (16.18)$$

where

T denotes the product or the item exposure or operational time in hours.

Figure 16.3.
General safety analysis approach.

Table 16.1. Classifications of Unsafe Event Effect.

Classification	Description	Value of α
I	safe	0
II	marginal	0.1
III	critical	0.5
IV	Severe or catastrophic	1

α denotes the unsafe event occurrence effect (this effect is categorized into four classifications as shown in Table 16.1).
F denotes the unsafe event frequency of occurrence.

EXAMPLE 16.7

The unsafe event occurrence frequency of an engineered system is 0.0005. The effect of the unsafe event occurrence is categorized as Class III (i.e., $\alpha = 0.5$). The estimated operational time of the system is 500 hours. Calculate the probability of an unsafe event occurrence.

Thus utilizing the above data in Equation (16.18) leads to:

$$P_{use} = (0.5)(500)(0.0005) = 0.1250$$

The probability of the unsafe event occurrence is 0.1250.

16.6   SUMMARY

This chapter briefly presents the historical aspects of reliability, maintainability, quality control and system safety along with their definitions.

Reliability measures, series and parallel networks, and reliability management are briefly discussed. Several numerical examples on the reliability topic are presented.

Important tasks of the maintainability management are described. These tasks are administration, coordination, documentation, design, and analysis.

Quality control is another area of the product assurance sciences. Therefore the chapter lists the important tasks of the quality control department and roles of the quality manager. In addition, the quality control manual is briefly described.

Lastly, the functions of system safety are listed and two procedures of system safety analysis are presented.

16.7   EXERCISES

1. Discuss briefly the history of reliability, maintainability, quality control, and system safety.
2. Describe at least six of the main responsibilities of a reliability engineering department.
3. Prove that the mean time to failure, MTTF, of a series network is given by

$$MTTF = \left[\sum_{i=1}^{n} \lambda_i \right]^{-1} \qquad (16.19)$$

where

n is the number of units in the series.
λ_i is the ith unit constant failure rate; for $i = 1,2,3, ---, n$.

4. Prove that the total failure rate (hazard rate), λ_s, of a series network is given by

$$\lambda_s = \sum_{i=1}^{n} \lambda_i \qquad (16.20)$$

The symbols n and λ_i are defined in exercise 3.

5. An independent unit parallel system (network) is composed of four non-identical units. At least one unit is needed to function successfully for system success. Units 1, 2, 3, and 4 probabilities of failure are 0.1, 0.15, 0.17 and 0.3, respectively. Calculate the system reliability.
6. Discuss the main functions of a quality control department.
7. Describe the main differences (if any) between the terms "quality assurrance" and "quality control."

16.8   REFERENCES

1. Rodgers, W. P. *Introduction to System Safety Engineering.* New York:Wiley Interscience (1971).
2. Dhillon, B. S. *Reliability Engineering in Systems Design and Operation.* New York:Van Nostrand Reinhold Company (1982).

3. Dhillon, B. S. and C. Singh. *Engineering Reliability: New Techniques and Applications.* New York:John Wiley & Sons Ltd. (1981).

4. "Maintainability Engineering Theory and Practice," *Engineering Design Handbook.* AMCP 706-133, Published by the Headquarters of U.S. Army Material Command, 5001 Eisenhower Ave., Alexandria, VA 22333.

5. Juran, J. M., F. M. Gryna and R. S. Bingham, eds. *Quality Control Handbook.* New York:McGraw-Hill Book Company (1974).

6. MIL-STD-882, "Systems Safety Program for System and Associated Subsystem and Equipment-Requirements for United States Department of Defense" (July 1962).

7. Nelson, R. S. "Integrating Reliability Progress into Design and Engineering: A Study of Management Problem," *Proc. of the Aerospace Reliability and Maintainability Conference.* pp. 178–184 (1963). Available from the IEEE.

8. Juran, J. M. and F. M. Gryna. *Quality Planning and Analysis: From Product Development Through Use.* New York:McGraw-Hill Book Company (1980).

CHAPTER 17

Engineering Maintenance Management

17.1 INTRODUCTION

Ever since the beginning of the Industrial Revolution, maintaining equipment in the field has been a challenging task. Since then, a significant amount of progress has been made to maintain equipment effectively in the field. As engineering equipment becomes sophisticated and expensive to produce and maintain, maintenance management has to face even more challenging situations to maintain effectively such items in operational environments.

In industry, the terms "maintenance" and "maintenance engineering" are frequently used. These terms have different meanings. The term "maintenance" means to keep equipment in operational condition or repair it to its operational mode. Both preventive and corrective activities are the components of maintenance. On the other hand, the "maintenance engineering" objective is to ascertain that the new products are designed for ease of performing maintenance and that the proper economic support subsystem is provided at the moment of need.

Thus, managing the functions of maintenance and maintenance engineering effectively is of great concern to present-day management. Therefore, this chapter explores various management oriented aspects of maintenance and maintenance engineering.

17.2 FUNCTIONS AND ORGANIZATION OF A MAINTENANCE ENGINEERING DEPARTMENT

The functions and organization of a maintenance engineering department are governed usually by factors such as industry, company size, thinking philosophy of management, etc. This section presents general guidelines on both these topics.

MAINTENANCE ENGINEERING DEPARTMENT FUNCTIONS

According to Reference [1], the following may be the major functions of a maintenance engineering department:

(i) Maintenance of installed equipment and facilities

(ii) Installation of new equipment and facilities

(iii) Modifications to already installed equipment and facilities

(iv) Inspection and lubrication of existing equipment

(v) Management of inventory

(vi) Disposal of waste and salvage

(vii) Production of utilities

(viii) Administration

(ix) Supervision of manpower

(x) Keeping records

ORGANIZATION OF MAINTENANCE ENGINEERING

Factors such as technical, geographical and management policy dictate the organizational structure of a maintenance organization. However, when one is organizing a maintenance engineering department one must take into consideration the following four points.

(i) Tailor the organization so that it can take into consideration the personalities involved and the changing conditions.

(ii) Division of authority must be defined so that it is concise and clear.

(iii) Optimum span of control (Generally, for the effective performance, a supervisory person should supervise between three and six individuals.)

(iv) Vertical lines of responsibility and authority must be designed so that they are as short as possible.

In regard to the reporting of the maintenance engineering department head, some people feel that the best results are obtained if he is directly responsible to top management. However, in some situations, the level of reporting for the maintenance engineering head does not effect the functioning of that department.

17.3 MAINTENANCE MANUAL

The maintenance manual is the bible for performing maintenance in an organization. It contains policies, objectives, procedures and organization concerning execution of the maintenance function. There are several reasons for having a maintenance manual in an organization. Some of them are unclear authority and responsibility, excessive bureaucracy, incomplete procedures, existence of duplicated effort and facilities, maintenance emphasis being given to unimportant areas, etc. The type and size of the manual may vary from company to company; however, according to Reference [1], before starting to write a manual, consideration must be given to the items shown in Figure 17.1.

The maintenance manual has various advantages and disadvantages. Some of the advantages are less paper work, reduction in cost of training manpower,

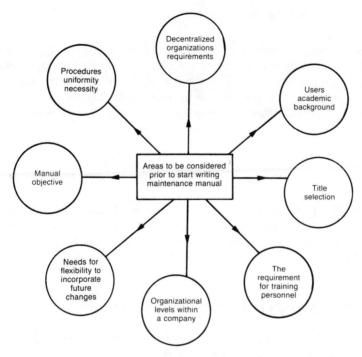

Figure 17.1.
Areas requiring examination prior to the writing phase of the maintenance manual.

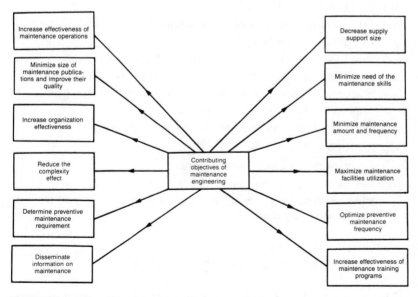

Figure 17.2.
Contributing objectives of maintenance engineering.

better management control, elimination of effort duplication, elimination of overlapping authorities and responsibilities, and increased understanding of the maintenance function. On the other hand, the disadvantages are the continual need for revision and updating to assure its effectiveness, impairment of innovations and improvements due to rigidity in practicing outline procedures, etc.

17.4   CONTRIBUTING OBJECTIVES OF MAINTENANCE ENGINEERING

Basically, maintenance engineering has the contributing objectives shown in Figure 17.2. These twelve objectives are self-explanatory, and therefore their details are not discussed. However, their descriptions may be found in Reference [2].

17.5   AN APPROACH FOR UPGRADING MAINTENANCE

This section presents an approach for managing maintenance functions systematically. The approach is composed of nine steps, as follows:

   (i) Determine the existing deficiencies. This can be established by interviewing the members of the maintenance organization as well as with the aid of the existing performance indicators.
  (ii) Set goals for maintenance improvement.
 (iii) Establish priorities by taking into consideration the savings from each project under study.
 (iv) Select indices to measure performance of each objective.
  (v) Establish plans for both short and long range. The short-range plan is accomplished within one year, whereas the long-range plan is completed within three to five years.
 (vi) Document both plans and forward copies to concerned bodies.
(vii) Implement plan.
(viii) Report status on a semi-annual basis.
 (ix) Review the plan at the end of each year and make necessary adjustments to the following year's short-range plan. Repeat the cycle of nine steps continuously.

17.6   EFFECTIVE MAINTENANCE MANAGEMENT

This section briefly discusses some of the important components of managing maintenance function. According to Reference [3], careful attention must be given to the following elements or components:

 (i) Maintenance policy
 (ii) Control of materials
 (iii) Preventive maintenance
 (iv) Work order
 (v) Job planning
 (vi) Priority and backlog control
 (vii) Data recording system
(viii) Performance measurement indices
 (ix) Work measurement

Items (i)–(viii) are briefly described in subsequent sections. The topic of work measurement is discussed in Chapter 20.

17.6.1 Maintenance Policy

For the smooth running of the maintenance function, each organization should have a document clearly describing the policies, objectives, authorities, responsibilities, maintenance performance measuring techniques, etc. In many companies this information is usually given in the maintenance manual. However, if a company has no such manual, then a document containing the essential information associated with the maintenance policy must be prepared.

17.6.2 Control of Materials

In the maintenance function, the spares and other materials have to be purchased, maintained in the inventory and delivered to the place of use. On the average, according to Reference [3], materials acount for 30 to 40 percent of the total direct maintenance costs. Thus the effectiveness of material control plays a key role in the efficiency of the maintenance organization. The improper control of materials may be the cause for the delay in performing maintenance, false starts, ineffective utilization of manpower, etc. In order to avoid these problems, it is mandatory to plan the maintenance job accurately, coordinate with puraching and stores, inspect the items received from outside sources and stock, coordinate the material delivery to the point of use, and inspect the accomplished job.

One of the most important problems of material control is deciding the carrying of spares in storage. When making such a decision, factors such as spare parts cost, acquisition lead time, part or equipment supported criticality and reliability, availability of the backup, and availability of spares in future must be considered. In addition, the usage of mathematical inventory control models should be given consideration. The diagram of the simplest model

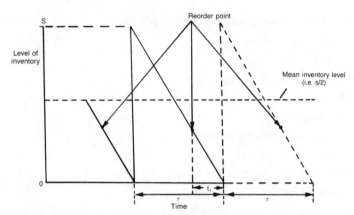

Figure 17.3.
Idealized situation diagram of an inventory control model.

used in inventory control is shown in Figure 17.3. One should note here that the model represents the idealized situation. Figure 17.3 shows that S number of spare units are ordered at the reorder point and are received at the exact time when the old stock is completely depleted. Thus the cycle continues. The comprehensive list of assumptions associated with the model may be found in Reference [4]. The following symbols were used to develop equations for the model:

S denotes the quantity of spare units ordered at one time.
M denotes the number of spare units required in one year.
H_c denotes the annual stock holding cost per spare unit.
P_c denotes the preparation cost of each order.
k_t denotes the total annual cost of inventory.
t_1 denotes the acquisition lead time.
τ denotes the cycle time.

Thus,

$$k_t = \text{Annual preparation cost} + \text{Annual stock holding cost}$$

$$= \frac{M}{S} \cdot P_c + \frac{S}{2} \cdot H_c \tag{17.1}$$

Differentiating Equation (17.1) with respect to S results in

$$\frac{d\,k_t}{dS} = \frac{H_c}{2} - \frac{M}{S^2} \cdot P_c \tag{17.2}$$

Thus, to find the optimal value of S, setting the derivative of Equation (17.2) equal to zero and solving for S yields

$$S^* = \sqrt{\frac{2\,M \cdot P_c}{H_c}} \qquad (17.3)$$

where S^* is the optimum quantity of spare units to be ordered at one time.
Utilizing Equation (17.3), the optimum time period, T^*, between orders is

$$T^* = S^*/M = \frac{1}{\theta^*} \qquad (17.4)$$

where θ^* is the optimum number of annual orders and is expressed as follows:

$$\theta^* \equiv M/S^* \qquad (17.5)$$

EXAMPLE 17.1

An engineering company uses 200 spare units annually for its equipment. The estimated values of H_c and P_c are \$50 per unit/year and \$100 per order, respectively. Calculate the optimum time period between orders. Assume that the spare units consumption decreases linearly and the units are instantly replenished when the stock is fully depleted.

Thus substituting the above specified data into Equation (17.3) yields

$$S^* = \sqrt{\frac{2(200)(100)}{50}} \cong 28 \text{ units}$$

Substituting the above result and the specified data for M into Equation (17.4) leads to

$$T^* = 28/200 = 0.14 \text{ years}$$

Thus, the optimum time period between orders is 0.14 years.

17.6.3   Preventive Maintenance

This is performed to keep equipment in satisfactory working condition by providing inspection and correction of early stage malfunctions.
The preventive maintenance need and scope are dictated by factors such as economics, standards compliance and process reliability. In the case of economics, the preventive maintenance cost is compared with downtime costs,

repair costs, etc., when making the preventive maintenance decision. Standards compliance means the government, insurance, professional, fire, safety and industral standards requirements for the periodic inspections. Lastly, process reliability takes into consideration the criticality of meeting production schedule deadlines, etc.

When making the preventive maintenance decision, one is faced with questions such as given below.

(i) What are the items that should be inspected in preventive maintenance?

(ii) What are the items which should be left out of the preventive maintenance program?

Answers to both these questions depend mainly on local situations. However, a good program should include, if applicable, items such as the equipment used for process, safety, utility, auxiliary, fire-protection, etc.

Before answering the second question one must consider the availability of the standby equipment, expected life of the equipment, cost of the downtime and repair, and the equipment criticality.

PREVENTIVE MAINTENANCE BENEFITS

Some of the main advantages of preventive maintenance are as follows [1]:

(i)   Reduction in catastrophic failures

(ii) Reduction in maintenance costs

(iii) Reduction in production downtime

(iv) Reduction in overtime payment for the maintenance workers

(v)   Reduction in the requirement of backup equipment

(vi) Reduction in the manufacturing unit cost

(vii) Improvement in safety for workers

17.6.4   Work Order

The purpose of the work order is to authorize and direct an individual or a group to carry out a specified task. A work order system should cover all the maintenance tasks requested and accomplished. The major benefits of the work order are that (i) it can be used by management to control costs, and (ii) it can be used for determining each job performance. The type and size of the work order may vary from one organization to another. However, a work order should be designed so that it includes at the minimum the following information concerning each job:

(i)   The requested and planned completion dates

(ii) The planned start date

(iii) Work description and its reasons

(iv) Cost of materials

(v) Cost of labour

(vi) Approval signatures

(vii) Work category (i.e., repair, installation, preventive maintenance, etc.)

(viii) Affected items

17.6.5 Job Planning

This is another element of effective maintenance management which is concerned with planning prior to the execution of a maintenance job. For example, in many cases, the following items have to be planned, coordinated and accomplished prior to the start of a maintenance job:

(i) Procurement of tools, components and materials

(ii) Securing safety work permit

(iii) Delivery of tools, components and materials to the maintenance work site

(iv) Methods and sequencing

(v) Development of designs

The degree of the job planning need largely depends on the local condition; however, generally it is accepted that on average for every 20 craftsmen there must be one job planner.

17.6.6 Priority System and Backlog Control

Normally in a maintenance organization it is not possible to perform maintenance as soon as it is requested. Therefore, the organization has to work according to some priority system. When assigning job priorities, factors such as the type of maintenance requirement, criticality of the item requiring maintenance, and required completion date of the job must be given careful consideration.

The backlog control is another element of effective maintenance management. In order to balance the work load and manpower needs, the job backlog must be identified. Furthermore, the information on backlog is needed to provide input to decisions related to manpower hiring and firing, maintenance work subcontracting, overtime, etc. Maintenance managers make use of various indices to control backlog.

17.6.7 Data Recording System

This plays an important role in the efficiency and effectiveness of the maintenance organization. The data recording system is used to keep various types of equipment or property records. Broadly speaking, these records can be classified into four categories. These are the maintenance costs, work performed, inventory, and files. The inventory record includes information such as equipment procurement date and cost, serial number, property number, size and type of equipment, location of the equipment, manufacturer, etc. On the other hand, the files are concerned with keeping the warranties, drawings, service manuals, operating manuals, etc.

All these records provide useful input into life cycle cost, reliability, maintainability and design studies. Furthermore, the information contained in records can be used as follows:

(i) When procuring new items

(ii) To determine operating performance trends

(iii) To make equipment replacement or modification decisions

(iv) To identify the areas of concern

(v) In investigating incidents

(vi) In troubleshooting breakdowns

17.6.8   Performance Measurement Indices

The main objective of all of these parameters is to indicate trends with the aid of past data as a reference point.

There are various indices which have been developed or used to measure the performance of the overall maintenance function or some individual elements of the effective maintenance management. Unfortunately, there is no one single index adequate for determining overall effectiveness of the maintenance function. However, several indices used in combination can serve this purpose.

This section presents a few selective indices; several more may be found in References [3] and [5].

INDEX I

This index is known as the broad indicator and is defined by

$$I_1 = C_{tmc}/P_0 \tag{17.6}$$

where

I_1 is the index parameter.
C_{tmc} is the total sum of maintenance costs.
P_0 is the total output. (Usually this output is given in units such as megawatts, gallons, tons, etc.)

Obviously, this index is used to indicate the overall performance of the maintenance function.

INDEX II

This is another index which belongs to the family of broad indicators. The index parameter, I_2, is given by

$$I_2 = C_{tmc}/S_t \qquad (17.7)$$

where S_t denotes the sum of total sales.

According to a study conducted a few years ago, the mean value of I_2 for all industry was equal to 5%. However, for the chemical and steel industries, the mean values of the index were 6.8% and 12.8%, respectively.

INDEX III

This is one of the indices used to control preventive maintenance function. The index parameter, I_3, is defined by

$$I_3 = T_{pm}/T_m \qquad (17.8)$$

where

T_{pm} is the total time spent in performing preventive maintenance.
T_m is the total time for performing overall maintenance.

According to past experience, the value of the index parameter should be controlled within 20% and 40% limits.

INDEX IV

This is another indicator used for determining preventive maintenance function and is defined by

$$I_4 = C_{pm}/C_r \qquad (17.9)$$

where

I_4 is the index parameter.
C_{pm} is the total cost of performing preventive maintenance.
C_r is the total cost of performing repairs.

17.7 SUMMARY

This chapter describes briefly the various aspects of maintenance management. The maintenance engineering organization and the functions of the maintenance engineering department are presented. The backbone of the effective maintenance organization, the maintenance manual, is described.

A procedure to upgrade maintenance and the contributing objectives of maintenance engineering are briefly discussed. Finally, the chapter goes on to describe the elements of effective maintenance management. These elements are maintenance policy, material control, preventive maintenance, work order, job planning, priority system and backlog control, data recording system, performance measurement indices, and work measurement.

17.8 EXERCISES

1. What are the major differences between the terms "maintenance" and "maintenance engineering"?

2. Describe the functions of a maintenance manager.

3. A maintenance organization consumes 400 parts of certain equipment per year. The annual stock holding cost per part is $90. Similarly, after analyzing various factors, the preparation cost of one order is estimated to be $200. It is assumed that parts consumption decreases linearly and that parts are instantly replenished when the inventory is fully depleted. Determine the optimum number of parts to be ordered at one time.

4. Describe at least four elements of effective maintenance management.

5. What are the advantages and disadvantages of the following:
   i) Maintenance manual
   ii) Preventive maintenance

6. What is the type of data which can be collected from the work order? Discuss the applications of such data.

7. Describe in detail at least five contributing objectives of maintenance engineering.

17.9 REFERENCES

1. Higgins, L. R. and L. C. Morrow, eds. *Maintenance Engineering Handbook*. New York:McGraw-Hill Book Company (1983).

2. AMCP 706-132. *Engineering Design Handbook: Maintenance Engineering Techniques*. 5001 Eisenhower Ave., Alexandria, VA 22333:Headquarters of United States Army Material Commmand (1975).

3. ER HQ-0004. *Maintenance Managers Guide*. Washington, D.C.:Energy Research and Development Administration (1976).

4. Riggs, J. L. *Production Systems: Planning, Analysis and Control*. New York:John Wiley & Sons (1981).

5. Dhillon, B. S. *Reliability Engineering in Systems Design and Operation*. New York:Van Nostrand Reinhold Company (1983).

CHAPTER 18

Introduction to Marketing

18.1 INTRODUCTION

Marketing is concerned with finding out the market needs and desires so that goods and services are made available to fulfill the needs and desires in question. The basic understanding of marketing is important not only to engineers and engineering managers but to every one of us. If the marketing process is accomplished efficiently, it will help to make everyone better off through higher standard of living, better life style, etc. The importance of marketing has increased during the past three and a half decades. In the future, it is also expected to grow.

Today, marketing produces and supports a vast number of jobs in North America because it is practiced by almost all business and non-business organizations. It makes us believe that a significant proportion of the end price paid for goods and services consists of costs for marketing.

This chapter explores the various aspects of marketing.

18.2 FUNCTIONS OF MARKETING

According to Reference [1], there are basically seven functions of marketing. Some articles and books may give more. For example, in Reference [2] it is explained that there are eight functions and in Reference [3] there are 47. Basically, these authors and writers are saying directly or indirectly the same thing. Therefore, in this section we will discuss only the seven functions of product marketing from Reference [1]. These are as follows:

(i) Product advertising
(ii) Product pricing
(iii) Collecting product market information
(iv) Product management
(v) Product physical distribution
(vi) Product distribution channel management
(vii) Personal selling

The above seven items are described briefly in the following sections.

18.2.1 Product Advertising

This is an important aspect of marketing which is concerned with informing the users or buyers regarding the availability of services and goods so they can purchase them to fulfill their needs and desires. In product advertising, the important duties which have to be performed include advertising budget determination, media selection for advertising, advertising message preparation, selecting agencies for advertising and advertisements effectiveness testing.

18.2.2 Product Pricing

This is another important function of marketing which deals with pricing products and services. People involved with pricing have to make many decisions in this regard. These include decisions on price objectives, price-setting basic procedure, if and when price changes should occur, developing discount policy, establishing a certain price for a product, etc.

18.2.3 Product Market Information (What is desired?!)

This information is provided by the marketing research unit to the decision makers in regard to market. Basically, this information pertains to finding out the needs of the market and the market satisfaction with the products and services developed and marketed to satisfy these needs. Thus the marketing research unit has to collect information on the following items:

  (i) Size of population in areas under consideration
 (ii) Population trends in the areas of concern
(iii) Size of potential customers and their location
 (iv) Size of the potential sales volume
  (v) Actual usage of the product by the market (in other words, how is it being used?)
 (vi) Buyers' habits
(vii) Identification of profitable and unprofitable segments of the market
(viii) Types of buyers
 (ix) Purchasing decision makers
  (x) Product price range
 (xi) Competition
(xii) Types of advertising
(xiii) Company image in the market

18.2.4 Product Management

The important responsibility of this function is to develop those new products and services which meet the requirements of the buyer. This task or responsibility is usually accomplished through the cooperation of manufacturing and research and development departments of the company. Various decisions are associated with each product marketed. For example, these decisions are concerned with types of packages, durability, raw materials, color, size, warranties, the brand name, product support in the field, etc.

18.2.5   Product Physical Distribution

This function deals with the manufactured goods movement from the manufacturing plant to the users. The manufactured goods transportation and storage are the two major items involved in this function. In the physical distribution area, the responsibilities to be accomplished are the determination of warehouse types, locations, sizes and number. In addition, the other duties associated with this function are making decisions regarding customer service, optimum number of items in the inventory, mode of transportation, shipment size, etc.

18.2.6   Product Distribution Channel Management

Generally the companies use middlemen such as wholesalers and retailers to market their produced goods. In this case the wholesalers and retailers are the distribution channels. The major duties concerned with this function are making decisions on whether or not to use middlemen, selecting middlemen, performing distribution channels effectiveness analysis, etc.

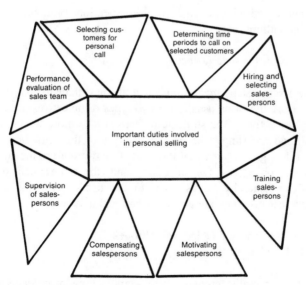

Figure 18.1.
Important duties involved in personal selling.

18.2.7   Personal Selling

This function is just like the advertising function. The only major difference between these two functions is that personal selling makes use of personal resources to inform customers about the existence of products which can satisfy their needs. In other words, the salesperson calls on buyers, customers, etc. To perform this function, there are several important responsibilities involved. Many of them are given in Figure 18.1.

18.3   MARKETING APPROACH

This section briefly presents a seven-step approach to making marketing decisions. These steps are normally accomplished to make the most of the marketing decisions. These steps are described in more detail in Reference [1] and are briefly discussed as follows.

18.3.1   Perform Market Analysis

This step is basically involved with identifying the needs of the market. Once these needs are known, the company can decide the type of market to be served, the best plan to serve the identified market, etc.

18.3.2   Perform Environmental Analysis

This is concerned with the analysis of competition, economy, technology, legal aspects, culture, etc.

18.3.3   Establish Feasible Objectives for Products

Once the first two steps are accomplished, the next thing to do is to establish objectives. Here the firms usually set two types of objectives, i.e., strategic and tactical objectives. The strategic objectives take more than one year to accomplish and are simply the basic objectives of the firm. Items such as volume of sales, growth, share of the market, and profit are the concerns of some strategic objectives. On the other hand, the achievement of tactical objectives can generally be assigned to the firm's marketing and other departments, because they are accomplished within a year.

18.3.4   Achieve Strategic Objectives Through
Product/Market Combinations

In this step, generally through several combinations of products and markets, the marketing department and management make decisions relating to

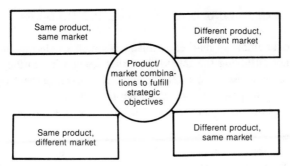

Figure 18.2.
Different product/market combinations when selling a product to a market for fulfilling strategic objectives.

the achievement of end strategic objectives. The four different product/market combinations are presented in Figure 18.2. These combinations are usually used to accomplish strategic objectives and are concerned with selling the product.

18.3.5 Achieve Tactical Objectives Through Marketing Mix

In Reference [1], the term "marketing mix" is defined as the marketing decision combination which is utilized to sell products over a certain time duration to certain markets. Therefore, this step is concerned with accomplishing tactical objectives by the means of marketing mix. Examples are improving the travelling time efficiency of the salespersons, switching the advertising media from the present one to another to increase the number of potential users or customers of the product in question, etc.

18.3.6 Achieve Objectives Through Marketing Organization

Once the decisions are made on how to accomplish the strategic and tactical objectives, then the next step is to set up the best marketing organization to achieve these objectives. In this case, one is concerned with determining the span of control, the basic organizational approach, degree of specialization, centralization/decentralization, the best procedure for motivating marketing persons, the size, type and number of subgroups in the marketing department, etc.

18.3.7 Establish a Feedback Control System

The objective of this step is to monitor the performance of the marketing department. Therefore, here one has to establish the monitoring aspect of the marketing operation, the measuring criterion, the performance standards, etc.

18.4   PRODUCT FAILURE DUE TO MARKETING MISTAKES

There are a number of marketing mistakes which are detrimental causes of a product not attracting the intended market. According to the results of a survey reported in Reference [2], in 1968 almost 80 percent of the 9,450 supermarket new products put into the market were unsuccessful. The following are the important reasons for the unsuccessful products in the market:

   (i) Introduction of products at the wrong time
   (ii) The firm entered into the wrong market. In other words, the company may have been quite successful in selling a specific type of product but its entry into different types of ventures is unsuccessful.
  (iii) The product's price is too high.
   (iv) Inadequate amount of effort put in to market the products
    (v) Product's unsatisfactory performance
   (vi) Product's differences are insignificant. In other words, the new product is not that much different from the successful product of another company.
  (vii) Limited market tests were not conducted.
 (viii) Poor quality

18.5   MATHEMATICAL MODELS USED IN MARKETING

This section presents various mathematical models used to make several types of marketing decisions. These models are as follows.

18.5.1   Model I

This model is due to Vidale and Wolfe [4] and can be used to make advertising decisions. The model is concerned with finding out the response of sales to advertising. Thus, from Reference [4], the equation of the model is:

$$R_s' = \alpha \, \beta[1 - (R_s)/(S_{st})] - \theta \, R_s \qquad (18.1)$$

where

the prime ($'$) on R_s denotes the differentiation with respect to time t.
S_{st} denotes the sales saturation level.
α denotes the rate of money spent for advertising at time t.
θ denotes the decay constant of sales. (The value of this constant is calculated when $\alpha = 0$ and is simply given by the fraction of sales lost per time unit.)

R_s denotes the rate of product sales at time t.

β denotes the constant of sales response. (The value of the constant is calculated when $R_s = 0$ and is equal to the sales generated per advertising dollar.)

R_s' denotes the change in R_s with respect to time t.

EXAMPLE 18.1

The data concerning a manufacturer's product is as follows:

$$\beta = 6, \; \theta = 0.15, \; \alpha = \$5,000, \; R_s = \$50,000,$$

$$S_{st} = \$125,000$$

Calculate the value of R_s.

Thus substituting the above given data into Equation (18.1) yields:

$$R_s' = (5000)(6)[1 - (50,000)/(125,000)] - (0.15)(50,000)$$

$$= 30,000 \, (1 - 0.4) - 7,500$$

$$= \$10,500$$

In this example, the manufacturer will sell $10,500 worth of additional product by spending $5,000 in advertising. It will be worthwhile to spend $5,000 in advertising only if the profit margin on $10,500 is above 50 percent.

18.5.2 MODEL II

This model is concerned with estimating the total market potential. In other words, it is used to find out the maximum amount of sales, K_s, of all companies in an industry, during a specified time duration, marketing effort, and environmental factors. According to Reference [5], the value of K_s may be estimated by utilizing the following equation:

$$K_s = \prod_{i=1}^{3} M_i \tag{18.2}$$

where

M_1 denotes the average price per unit.

M_2 denotes the total number of customers of a certain product under the specified conditions.

M_3 denotes the expected number of units bought by an average customer.

EXAMPLE 18.2

For a particular type of product, there are about 75,000,000 customers per year. Each customer on average purchases five units per year. Each unit mean price is $20. Calculate the value of the total market potential.

In this example the values of M_1, M_2, and M_3 are equal to 20, 75,000,000, and 5, respectively. Thus utilizing the specified data in Equation (18.2) results in

$$K_s = (20) \cdot (75,000,000)(5)$$

$$= \$7,500,000,000$$

18.5.3   Model III

This model is used to evaluate the buying behavior of the consumer and is known in published literature as either the expectancy-value model, the multi-attribute attitude model, or the linear compensatory model. The model takes into consideration the customer attitude toward each brand of the product and the weight of importance customers give to each attribute of the brand. Thus, from References [5] and [6], the equation of the model is

$$N_{ij} = \sum_{k=1}^{m} w_{kj} M_{kij} \qquad (18.3)$$

where

m denotes the number of attributes of a specified brand. (These attributes are important in the selection of the brand in question.)

N_{ij} denotes the attitude score of customer j for brand i of the product in question.

M_{kij} denotes the amount of attribute k due to product brand i because of customer j's belief.

w_{kj} denotes the weight of importance given to product attribute k by customer j.

EXAMPLE 18.3

A customer has to choose from two brands of an engineering product. Furthermore, the customer considers that there are only two attributes, A and B, of the product which are important to him. In the buyer's opinion, attribute A is two times more important than attribute B. For both brands, the given scores, out of a maximum of 10, for attributes A and B were as follows:

Product Brands	Attribute A	Attribute B
x	7	10
y	6	9

Predict the customer attitude toward both brands.

In the above example, the data is specified for the following items:

$$m = 2, \ M_{Ax1} = 7, \ M_{Ay1} = 6, \ M_{Bx1} = 10, \ M_{By1} = 9$$

$$W_{A1} = 2, \ W_{B1} = 1$$

Thus substituting the above data into Equation (18.3) yields:

$$N_{x1} = (w_{A1}) \ (M_{Ax1}) + (w_{B1}) \ (M_{Bx1}) = (2)(7) + (1)(10) = 24$$

$$N_{y1} = (w_{A1}) \ (M_{Ay1}) + (w_{B1}) \ (M_{By1}) = (2)(6) + (1)(9) = 21$$

In this case the consumer is more attracted towards product brand x.

18.5.4   Model IV

This is simply an index used to predict, for a specified area, the percent-age, α, of total national buying power. In Reference [5], this index is defined as follows:

$$\alpha = 0.2 \ i + 0.5 \ j + 0.3 \ k \tag{18.4}$$

where

i denotes the proportion of the total national population residing in a speci-fied area.

j denotes the proportion of the total national disposable personal income in a specified area.

k denotes the proportion of the total national retail sales in a specified area.

EXAMPLE 18.4

For a certain city, the value of i, j and k are as follows:

$$i = 0.2 \text{ percent}$$

$$j = 0.5 \text{ percent}$$
$$k = 0.4 \text{ percent}$$

Calculate the value of the buying-power index for that city. Utilizing the given data in Equation (18.4) yields

$$\alpha = 0.2(0.2) + 0.5(0.5) + 0.3(0.4)$$

$$= 0.41 \text{ percent}$$

Thus the value of the buying-power index for the city is 0.41 percent. Therefore, one may say that 0.41 percent of the country's sale might take place in that city.

18.5.5   Model V

This model is concerned with computing the total profits. The profit equation from Reference [5] is

$$Y = x\, S_u - k\, S_u - B \tag{18.5}$$

where

$$x = p - A_\ell \tag{18.6}$$

$$B = c_f + c_m \tag{18.7}$$

The identifications of the symbols used in Equation (18.5) are as follows:

c_m denotes the discretionary costs of marketing.
S_u denotes the total number of product units sold.
c_f denotes the total fixed costs.
y denotes the total profits.
k denotes the costs of manufacturing and distribution.
A_ℓ denotes the cost of allowance for each unit. (This cost includes discounts, commissions, etc.)
p denotes the list price of the product.

Thus the total profit may be calculated by utilizing Equation (18.5).

18.5.6   Model VI

This model deals with calculating the total cost of a proposed product distribution system. Thus, the total distribution cost, k_t, is given by

$$k_t = \sum_{i=1}^{4} k_i \qquad (18.8)$$

where

k_1 denotes the lost sales total cost because of the proposed distribution system.

k_2 denotes the proposed distribution system total freight cost.

k_3 denotes the proposed distribution system total fixed cost of warehousing.

k_4 denotes the proposed distribution system total variable cost of warehousing.

18.5.7 Model VII

This model is concerned with finding the value of a prospect. This value, M, from Reference [5] is given by

$$M = y\ P(k) - Z_1 \qquad (18.9)$$

where

y denotes the present worth of the income expected from a new buyer (customer) in future.

$P(k)$ denotes the probability of prospect conversion and the function of k, the number of calls made to convert the prospect into a buyer (customer).

Z_1 denotes the amount of investment required for the conversion of prospect to buyer (customer).

The following formula yields the value of y:

$$y = \sum_{j=1}^{n} (1 + i)^{-j} (G_m S_{ej} - C_c) \qquad (18.10)$$

where

n denotes the time in years. (This is that time in which the new customer is expected to remain a customer of the company.)

G_m denotes the gross margin on sales.

i denotes the discount rate.

C_c denotes the annual cost of maintaining contact with the customer.

S_{ej} denotes the expected sales in year j from the new customer.

Similarly, the value of the Z_1 can be obtained from Equation (18.11):

$$Z_1 = \alpha \, k \qquad (18.11)$$

where α denotes the cost of a call made to convert the prospect into a customer.

18.6 SUMMARY

The chapter briefly summarizes the important basics of marketing which will be useful to engineers to grasp the concept of marketing.

The marketing functions such as advertising, pricing, market information, product management, physical distribution, distribution channel management and personal selling are briefly explored. A seven-step marketing approach is discussed. The pertinent marketing mistakes which are causes of product failure are briefly described. The last portion of the chapter presents seven mathematical models which have applications in marketing.

18.7 EXERCISES

1. Write an essay on the historical aspect of marketing.
2. What are the essential elements of marketing?
3. What are the main differences between functions of an engineering manager and a marketing manager? In addition, discuss their similar functions.
4. Discuss the recommendations for avoiding new engineering product failure in the market.
5. What are the essential components of the term "personal selling" used in marketing?

18.8 REFERENCES

1. Hise, R. T., P. L. Gillett and J. K. Ryans. *Basic Marketing: Concepts.* Cambridge, MA:Winthrop Publishers, Inc. (1979).
2. Diamond, J. and G. Pintel. *Principles of Marketing.* Englewood Cliffs, NJ:Prentice-Hall, Inc. (1972).
3. Turck, F. B. "The Forty-Seven Functions of Marketing," in *Management Guide for Engineers and Technical Administrators.* N. P. Chironis, ed. New York:McGraw-Hill Book Company, pp. 300–303 (1969).

4. Vidale, M. L. and H. B. Wolfe. "An Operations-Research Study of Sales Response to Advertising," *Operations Research*, 370–381 (June 1957).

5. Kotler, P. *Marketing Management: Analysis, Planning and Control*. Englewood Cliffs, NJ 07632:Prentice-Hall, Inc. (1980).

6. Wilkie, W. L. and E. A. Pessemier. "Issues in Marketing's Use of Multi-Attribute Attitude Models," *Journal of Marketing Research*, 428–441 (1973).

CHAPTER 19

Product Warranties and Liabilities

19.1 INTRODUCTION

In recent years, considerable attention has been given to product warranties and liabilities. For example, the United States Congress has passed several pieces of legislation in the liabilities area. This has expanded the manufacturers' duties and obligations toward their products. Therefore it may be said that today if anyone is injured because of a defect in the product, then the manufacturer must consider himself to be liable under the law of the country. According to Reference [1], the significant consumer-initiated lawsuits in 1976 were on the order of 50,000. This estimate was presented in Senate testimony by consumer advocate Ralph Nader. However, according to the estimate of the Defense Research Institute, the products cases in litigation were 100,000 in 1966. This figure increased to over 500,000 per annum in the early seventies. It was predicted by the Institute that the lawsuits number would reach 1,000,000 by 1980.

During the sixties [2], a form of contract known as warranty had been practiced by the Department of Defense, commercial aircraft manufacturing companies, and airline companies. Nowadays it is usual that the product manufacturers provide some sort of written documents which guarantee the integrity of their products to the buyers. These documents outline the responsibilities of the companies if the manufactured products turn out to be defective.

According to Reference [3], out of 369 U.S. manufacturers examined by the Conference Board, only one percent of them had no documented or written warranties on their products. Furthermore, the survey revealed that the average cost of a warranty claim was less than the 2% of the sales value of a product. In addition, the manufacturers cited the three basic reasons for providing warranties on their manufactured goods. These reasons were providing support for their manufactured goods, defining liabilities and responsibilities, and competition.

Both warranties and liabilities are described in the following sections of the chapter.

323

19.2 WARRANTIES

This section presents the various areas concerned with product warranties. The material for some portions of this section is based on the findings in Reference [3]. The Reference [3] report is the result of an examination of 369 U.S. manufacturing companies.

19.2.1 Warranty Primary Obligations

The manufacturers' obligations for electrical and mechanical goods is the basic concern of this section. Therefore, in general, options opened to electrical and mechanical product manufacturers to fulfill their guaranteed warranty obligations are as follows:

(i) Replace the defective product.

(ii) Replace the faulty components in the product or pay the purchasers so that they can procure such components.

(iii) Make payments for the service labor to rectify defective components or to replace defective components with new ones.

(iv) Make payments for the replacement of defective parts and the installation or repair cost of such parts.

According to Reference [3], the majority of manufacturers surveyed indicated that during the warranty period they pay for both the replacement parts and the service labour. This warranty period normally extends for one year.

19.2.2 Reasons for the Warranty Needs

This section summarizes the reasons for the warranty requirement. According to Reference [4], some of the reasons for having a warranty are as follows:

(i) It makes the producer of the product responsible for rectifying faults in his product.

(ii) It assures the buyer that the manufactured product will at least satisfy his contractual specifications.

(iii) It helps in prompt acceptance of the produced product.

(iv) It helps to enhance the manufactured product marketability.

(v) It encourages better monitoring of the performance of the product.

(vi) It helps to expedite payment from the customer for the sold product.

(vii) It provides a means of assuring that the technical services will be at hand at the moment of need..

(viii) It encourages the manufacturer to produce better product.

(ix) During the no-charge warranty period, it allows the customer to gain proficiency in operating the product.

(x) It forces the product manufacturer to provide better product support services and documentation.

19.2.3 Variation in Warranty Practices for Export Products

According to the survey conducted in Reference [3], nine out of ten American manufacturers stated that some of their produced goods are sold in foreign markets. About 75% of the manufacturers indicated that their warranty terms are the same for both foreign and domestic markets. However, the remaining 25% said that they have different warranty terms for the product to be sold in other countries. These companies have cited the following reasons for developing different terms:

(i) The product warranty duration period may depend on factors such as service quality, reliability and maintenance availability in local environments.

(ii) Heavy reliance on local dealers and distributors where the product is sold because it may present an obstacle to manufacturers to properly fulfill their warranty obligations as they would have done in the domestic market.

(iii) Availability of information on the usage of product may be less in the foreign market.

(iv) Different legal jurisdictions of the outside countries may demand changes in the warranty terms.

Finally, the other three important points to be noted concerning the goods produced for the export market are as follows:

(i) Usually the warranty period is shorter for products to be sold in foreign countries as compared with products for the domestic market.

(ii) The defective product components are generally replaced with new components or the entire faulty product is replaced with a new one.

(iii) The extended warranty coverage option to the customers in outside countries is less likely to be offered.

19.2.4 Warranty Management

This section briefly discusses the organization of the warranty administration function. According to Reference [3], in the majority of manufacturing companies the warranty administration responsibility is generally assigned to either product or technical service functions. Only a small percentage of manufacturers have a separate unit to handle warranties. However, there are many factors which dictate the positioning of the warranty administration responsibility in the organizational structure of a company. For example, the organizational structure of the firm, the firm's distribution system characteristics, the type of product sold, customer's importance in warranty-protection issues, frequency and type of claims are among the primary factors. Firms may assign responsibility for warranty administration to:

(i) customer service or sales unit when the claims are relatively simple and their number is low.

(ii) product service or technical service function when the claims require a certain technical knowledge to process.

(iii) a sole separate warranty administration unit when the cost or number of claims is high.

19.2.5   Trends in Warranty Claims

A survey of manufacturers [3] conducted in recent years shows that the requests for warranty repairs or service during the previous two years were as follows:

(i) Fifty-one percent of the companies reported that there was no change in the number of warranty related requests.

(ii) Thirty-eight percent of the manufacturers agreed that the requests had increased.

(iii) Only eleven percent of the participating organizations indicated that warranty repair requests had decreased.

From the above findings one can draw his own conclusion. The following are the common reasons given for the growth in warranty claims:

(i) Greater product volume increases the chances for more warranty claims.

(ii) Product misuse by the customer

(iii) Inflationary pressures

(iv) Deterioration in the quality of manufactured goods

(v) More complex and service-sensitive products

(vi) Poor maintenance in the customer facilities due to poorly trained and unqualified maintenance personnel

(vii) More customer awareness of manufacturers' warranty obligations

The manufacturer service executives forecasted the following warranty problems in the future:

(i) Government regulations

(ii) Rapid increase in the cost of material, labor, etc.

(iii) Training users and dealers for the proper usage of product

(iv) Increasing trend in user expectations

(v) Warranty starting dates records-keeping

(vi) Rising litigation activity

(vii) Shortage of qualified product service manpower

(viii) Rising product liability

19.2.6 Reliability Improvement Warranty

In the last decade considerable attention has been given to reliability improvement warranty (RIW). It is concerned with encouraging the equipment manufacturers with incentives to produce more reliable products with less field repair costs. In the past, the U.S. Department of Defense has been the prime user of this approach when procuring new equipment. Therefore, this section discusses various areas which are related to RIW.

COMPONENTS OF A WARRANTY

Components of a warranty are to be designed in such a way that the requirements in question are fully satisfied. For example, in the airline business the

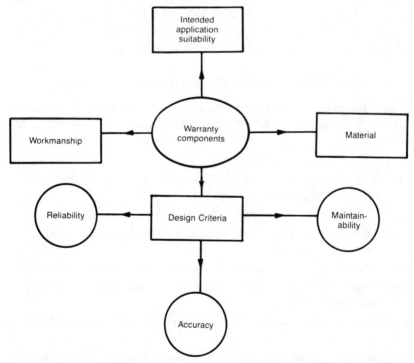

Figure 19.1.
Elements of a warranty in the airline business.

warranty elements [4] as shown in Figure 19.1 are made up of design criteria, material, workmanship, and suitability for satisfying intended application.

The design criteria are concerned with reliability, maintainability and accuracy. The items considered in reliability are mean time between failure, time between overhauls, expected life, number of landings, etc. On the other hand, maintainability takes into consideration measures such as mean time to repair, maximum material cost, mean material cost per repair, maximum cost per landing and per operating hour: maximum cost for aircraft, engine, etc.

FACTORS CONCERNING WARRANTY ELEMENTS

This section presents the factors which must be considered by the buyer when specifying elements of warranty. Therefore, according to Reference [4], the customer must probe into the product producibility, the product line complexity, measurability of warranty components, management capability of manufacturer, product line planned longevity, maturity with manufacturer, manufacturer financial health, technical support capability of the manufacturer, overall support availability of the manufacturer, etc. It is to be noted here that these factors are outlined in Reference [4] for the airline customer.

However, from the manufacturer side, when specifying elements of warranty one must take into consideration the warranty cost, accessibility to user records, type of warranty desired by the user, user capability, training plans, financial health, operational and maintenance plans, etc.

MANUFACTURER'S RELIABILITY IMPROVEMENT WARRANTY RISKS

According to Reference [5], there are several risks to the manufacturer associated with the product reliability improvement warranty. These are as follows:

(i) Sudden increase in the inflation rate

(ii) Due to competitors' pressure, the manufacturer may bid too low a price for the product.

(iii) Covering a high technology product

(iv) Unforeseen operational stresses under which the manufactured product has to function

(v) Product exposure to failure will increase due to increase in its usage.

(vi) Due to maintenance manpower, product mishandling or tampering may induce failures.

(vii) Inability of the manufacturer to evaluate precisely the product failure frequency

(viii) Inability of the manufacturer to determine accurately the product repair cost

CRITERIA FOR SELECTING EQUIPMENT FOR WARRANTY COVERAGE

According to Reference [6], when selecting equipment for warranty coverage, the following factors must be considered:

(i) Ready transportability of the equipment
(ii) Precise knowledge of product usage environment and the expected operation time
(iii) Self-containability of the equipment
(iv) Field-testability of the product
(v) Initial support cost of the product is moderate to high.
(vi) Development of the equipment to the stage where its reliability and maintainability can be evaluated with a certain degree of accuracy

INCENTIVES OF WARRANTY TO BOTH MANUFACTURER AND CUSTOMER

The warranty is beneficial not only to the customer but to the manufacturer as well. Thus, the primary incentives to manufacturer and user are as follows.

MANUFACTURER

(i) Improvement in chances for more business in the future
(ii) Expanded market for the product
(iii) Improvement in chances for repeat business
(iv) Cost controls establishment through product design, production and field support

USER

(i) Decreased risk
(ii) Low initial and operational costs mean better cost versus profit ratio.
(iii) State-of-the-art advancement: reliability, maintainability, accuracy, operation and function
(iv) Calculable risk

WARRANTY COST MODEL

In recent years, there have been various newly developed mathematical warranty cost and price models. Some of these may be found in References [7] and [8]. This section presents a model taken from Reference [5] to estimate contractor warranty cost, C_w, as follows:

$$C_w = c_f + \lambda \{Hc_r\} \tag{19.1}$$

where

λ denotes, for all warranted equipment, the constant failure rate per hour during the specified warranty period.

H denotes, for all warranted equipment, the total operating time in hours over the specified warranty period.

c_r denotes the average cost to the manufacturer to repair a warranted piece of equipment.

c_f denotes the warranty fixed costs to the manufacturer.

19.3 PRODUCT LIABILITIES

Due to government legislation, an increasing number of law suits, and liability insurance rate increases, considerable importance is being given to product liabilities. Therefore, this section briefly discusses the various aspects of product liabilities.

19.3.1 Causes for Liability

This section briefly lists the causes for the liability suit. Thus, according to Reference [1], a product manufacturer in the U.S.A. may be liable for:

(i)   Failure to keep suitable records of user complaints and product failures

(ii) Design defect in the manufactured product. In other words, it is not suitable for its specified purpose.

(iii) The product is not adequately labeled for correct use and possible warnings.

(iv) Due to inadequate testing and quality control, the product has manufacturing defects.

(v)   Failure to establish adequate product manufacturing, distribution and sales records

(vi) The packaged product is prone to handling and safety-related damage.

(vii) The newly manufactured product is packaged so that its components can be sold individually in the market.

(viii) The packaged product is being sold in an incomplete and dangerous form.

(ix) The instructions for the packaged product can become detached before it is sold, thus allowing the sale of that product in a dangerous and incomplete form.

19.3.2   Areas Examined in Building a Product Liability Suit

To build a product liability suit, several areas are probed. Some of them are as follows:

 (i) Incorrectness of the product safety procedure
 (ii) Quality of the material used to produce the product
 (iii) Test procedures
 (iv) Inadequacy in design
 (v) Faults in design
 (vi) Departure from set product manufacturing procedures, design specifications, etc.
 (v) False claims made in product advertisements

However, in the last few years, liability suits have already established the negligence of items such as incorrect testing to find faults, incomplete or misleading instructions for the product operation, failure to provide warning of possible dangers to users, etc.

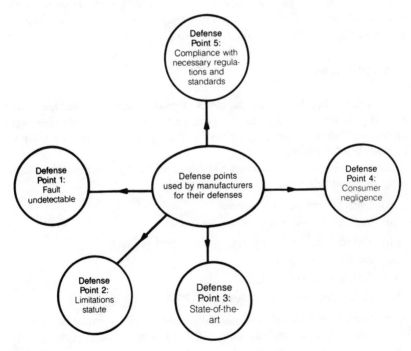

Figure 19.2.
Defense points used by manufacturers for their defense.

19.3.3   Defenses Used Against Consumers by the Suppliers

When a user of a product charges negligence on the part of the manufacturing company there are basically five points which may be used for the defense [1]. These are shown in Figure 19.2.

The first defense point is concerned with "fault undetectability." Here the manufacturer pleads that no technique presently exists to test for the fault which resulted in failure and injury. Therefore, the manufacturing company did its best by giving special care to the design and marketing aspects.

The second defense point is concerned with the "statute of limitations." Here, questions such as "When did the accident concerning the product happen?" and "When did the damages due to the accident manifest themselves?" are asked.

The state of the art in design is the third defense point. With this the manufacturer defends himself by stating that the product is designed as safe as possible. In other words, it could not be designed safer by anyone else.

The fourth point which may be used for the manufacturer's defense is concerned with consumer negligence. Here the manufacturer defends by saying that the user used the faulty product voluntarily and knowingly.

The fifth and last defense point is concerned with compliance with applicable standards. This defense point simply states that the manufactured product complies with the applicable government regulations, standards, etc.

19.4   SUMMARY

This chapter briefly presents the topics of product warranties and liabilities. Basically the chapter is divided into three parts, which are introduction, warranties and liabilities. In the introduction, the topics of warranties and liabilities are briefly introduced.

The topic of warranties is covered in fair depth. The following are the main components of this topic:

(i)   Warranty primary obligations

(ii)   Reasons for the warranty needs

(iii)   Variation in warranty practices for export products

(iv)   Warranty management

(v)   Trends in warranty claims

(vi)   Reliability improvement warranty (RIW)

The items covered under the reliability improvement warranty section are the warranty elements, manufacturer's RIW risks, selection of equipment for warranty, warranty incentives, etc.

Product liabilities is the other topic of concern. Here attention is directed to liability causes, areas for building a product liability suit, and defense points used against consumers by the manufacturers. At the end, the chapter lists the relevant references on the topics under discussion.

19.5   EXERCISES

1. Write an essay on product liability suits.
2. What are the advantages to the users of having a product warranty program?
3. What are the benefits and the drawbacks to the manufacturer of having a product warranty program?
4. Are the warranty requirements different for the export market products in comparison to the ones for the domestic market? If yes, what are the reasons?
5. What are the reasons for the growth in warranty claims?
6. What is meant by the term "Reliability Improvement Warranty"? List the elements of a warranty.
7. What are the significant causes for which a product manufacturer may be liable?

19.6   REFERENCES

1. Kolb, J. and S. S. Ross. *Product Safety and Liability.* New York:McGraw-Hill Book Company (1980).
2. Bonner, W. J. "Warranty Contract Impact on Product Liability," *Proceedings of the Annual Reliability and Maintainability Symposium.* pp. 261–263 (1977).
3. McGuire, E. Patrick. "Industrial Product Warranties: Policies and Practices," A Research Report from the Conference Board, available from the Conference Board, Inc., 845 Third Avenue, New York, NY 10022 (1980).
4. Flottman, W. W. and M. R. Worstell. "Mutual Development, Application and Control of Suppliers Warranties, American Airlines and Litton Systems-Aero Products Division," *Proceedings of the Annual Reliability and Maintainability Symposium.* pp. 213–221 (1977).
5. Balaban, H. S. and M. A. Meth. "Contractor Risk Associated with Reliability Improvement Warranty," *Proceedings of the Annual Reliability and Maintainability Symposium.* pp. 123–129 (1978).
6. Balaban, H. and R. Retterer. "The Use of Warranties for Defense Avionics Procurement," *Proceedings of the Annual Reliability and Maintainability Symposium.* pp. 363–368 (1974).
7. Gates, R. K., R. S. Bicknell and J. E. Bortz. "Quantitative Models Used in the RIW Decision Process," *Proceedings of the Annual Reliability and Maintainability Symposium.* pp. 229–236 (1977).
8. Balaban, H. and R. Retterer. "Guidelines for Application of Warranties to Air Force Electronic Systems," ARINC Research Corporation RADC Report TR-76-32 (March 1978).

CHAPTER 20

Introduction to Work Study

20.1 INTRODUCTION

In these inflationary times, due to the increasing cost of producing products and other factors, the topic of work study has been receiving increasing attention. Therefore, this chapter is devoted to the subject of work study. Work study, according to Reference [1], may simply be referred to as a study of performance. In addition, it is applied to generate measurement standards and to enhance efficiency at the workplace. The components of work study are the method study, work measurement and incentive wage. Method study deals with investigating the currently practiced approaches to performing work in order to find room for further development.

On the other hand, the work measurement objective is to determine the amount of time taken by a worker to perform a certain task. The last element of work study, the "incentive wage," is intended for workers to increase their motivation and productivity. In other words, it gives a proportionately higher wage for their better effort.

The various aspects of work study are discussed in the following sections.

20.2 REASONS FOR AND CRITICISMS OF PERFORMING WORK STUDY

This section briefly discusses the reasons for management to apply the work study concept, as well as criticisms against conducting such a study.

20.2.1 Reasons for Conducting Work Study, from the Management Standpoint

According to Reference [2], some of the important reasons for management to apply the work study concept are as follows:

(i) It can be applied in different situations where manually oriented work tasks are being performed, for example, in factories, laboratories, offices, restaurants, banks, etc.

(ii) It is a systematic concept.

(iii) It is the most useful and precise approach to establishing performance standards.

(iv) It is a relatively inexpensive approach to increase productivity.

(v) It is the method, if used correctly, that leads to instant savings. Furthermore, these savings continue as long as the improved version of the operation is practiced.

(vi) It is an invaluable tool available to management for attacking inefficiency in an organization.

20.2.2 Work Study Criticisms

The following are the common criticisms of work study [1]:

(i) Requires more supervision

(ii) Leads to poor relationships between management and worker

(iii) Leads to increase in administrative responsibilities

(iv) Inadequate consideration for factors such as loneliness, monotony, environmental conditions, special clothing, etc.

(v) Imposes safety problems

20.3 BASIC APPROACH TO PERFORMING WORK STUDY

Different work study practicioners may define their own steps to perform work study. However, Reference [2] presents an eight-step approach which is shown in Figure 20.1.

20.4 WORK STUDY DEPARTMENT AND ENGINEER

The objective of this section is two-fold: a) to describe briefly the work study department; b) to discuss the qualities of the work study engineer. Both these items are discussed separately as follows.

20.4.1 Work Study Department

A well-established work study department is expected to perform duties such as calculating time standards, work measurement, method studies, developing incentive plans for workers, evaluating the proposed work changes, etc.

In regard to the position of the work study department, it must be remembered that work study is a service to the company management and therefore must be treated as a staff activity, rather than a line activity. Different companies have different titles for the head of the work study department and his

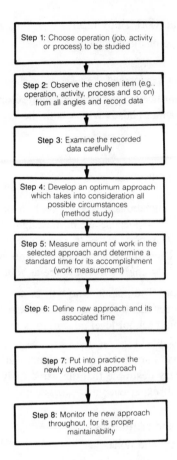

Figure 20.1.
A procedure for conducting work study.

reporting position. However, the one important thing that must always be considered is that such a person must be placed so that his sound recommendations are acted upon without any difficulty. Generally, it is understood that the role of the work study department head is mainly of an advisory nature. Therefore, he normally occupies a high position in the management structure. In small companies the work study department head should report directly to the president or a senior executive in order to make sure his recommendations are taken into consideration, because in small companies, generally, the decisions made by the president or the senior executive are the ones which really count.

Usually there are the following factors considered when deciding the position of the head of the work study department in an organization:

(i) Qualifications and experience of the person in question

(ii) Personality of the work study head

(iii) The size and type of the company

(iv) The industry

(v)   The type of work performed

(vi) Other department heads' qualifications and experiences. These are only those department heads with whom the work study head has to interact.

20.4.2 Work Study Engineer

There are different designations of a work study man in different organizations. Nevertheless, for our purpose we have designated for him the title of work study engineer. This person has to possess certain qualifications and qualities for success. Therefore this section presents the qualifications and qualities of the work study engineer. These are as follows:

(i)   Formal Education: The person who will be trained as a work study man or an engineer has to have a certain level of education. Generally, the high school level education is the acceptable minimum requirement.

(ii) Experience: This is another important qualification of the man in question. It is very desirable that the work study person has some practical (manual) experience as a worker in jobs which he will be monitoring.

(iii) Self-confidence and tactfulness

(iv) Empathy, sincerity and honesty

(v)   Enthusiasm: This is important for his job in order to transmit enthusiasm to others who work along with him.

(vi) Good appearance: This means the person must appear efficient, tidy, neat, etc.

20.5 METHOD STUDY

This is concerned with the analysis of currently used methods to perform work tasks in order to find latitude for further improvement. The analysis is accomplished through a two-step process as follows:

(i)   Breaking down each work operation into various principle activities

(ii) Examining each of these activities carefully to locate faults in the present ways of performing work

When selecting jobs for the method study investigation, human reactions and economic and technical factors must be considered. Various elements of the method study are discussed in the following sections.

20.5.1 Application Areas of Method Study

The method study concept is applied to a wide range of problems. The typical types of jobs studied are as follows:

  (i) Layout of the factory from the standpoint of movement of labour force
 (ii) Materials handling
(iii) Layout of the factory from the movement of materials standpoint
 (iv) Layout of the workplace
  (v) Manufacturing sequence from one end to another
 (vi) Operation of automatic machines
(vii) Workers' movements at the place of work

20.5.2 Methods Engineering Techniques

According to Reference [3], the following are the important techniques of methods engineering:

  (i) Process charts
 (ii) Operation analysis
(iii) Work measurement
 (iv) Work sampling
  (v) Motion study
 (vi) Value analysis

In this section we will discuss only the process charts and operation analysis, while work measurement will be discussed in a separate section to follow. Most of the remaining techniques are described in References [1–5].

PROCESS CHARTS

These charts are used to present graphically the events occurring in a series of operations. In the graphical form these events are easy to analyze. The activities occurring during a process are categorized into five different classifications such as:

  (i) *Operations:* These signify the pertinent steps in a process. Broadly speaking, during an operation the material, component or product under consideration is transformed, changed or modified. The symbol used to represent an operation is:

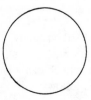

(ii) *Inspection:* This is concerned with the quality and correctness of the subject under consideration. The following symbol is used to represent inspection:

(iii) *Transportation:* This is concerned with the movement of the subject of the study (e.g., workers, materials, or equipment) from one point to another. Transportation activity is denoted by the following symbol:

(iv) *Storage:* This activity is concerned with holding an item under monitored conditions. The following symbol is used to represent this activity:

(v) *Delay:* This means that there is an interruption between two consecutive operations. This is denoted by the following symbol:

In addition, activities performed simultaneously or by the same worker at the same work point (for example, operation and inspection performed simultaneously) are denoted by the following combined symbol:

All of the above symbols were recommended by the American Society of Mechanical Engineers. These symbols are utilized to develop process charts in Reference [2].

There are basically three types of process charts. These are as follows:

(i) *Operation Process Charts:* These charts are used to show the operations and inspection accomplished in a process. Simplicity is the greatest advantage of these charts, and their usage helps in cost reduction through combining and eliminating specific activities.

(ii) *Flow Process Charts:* The major difference between these and operation process charts is the inclusion of material handling and storage activities in the flow process charts. These charts' significant advantage is that they depict material handling operations graphically. Furthermore, the material handling operations comprise a significant component of most product costs.

(iii) *Multiple Activity Charts:* These are also known as the man and machine charts. These charts present the active and idle time of two or more machines, men or a combination of both (i.e., men and machines).

OPERATION ANALYSIS TECHNIQUE

This is another important technique of method study engineering. This approach is utilized to investigate the areas which affect the procedure to accomplish an operation at lowest cost. Thus the technique is concerned with applying the questioning attitude to individual components of an operation. In this technique, the individual points of primary analysis are examined carefully. These points are the work environments, material, process analysis, handling of material, method, inspection needs, operation objective, part design, workplace layout, common job improvement areas and tool equipment. Past experience indicates that through the application of operation analysis technique on almost all occasions there will be room for methods improvement.

20.5.3 Method Study Advantages

There are a large number of advantages of the method study. In relation to construction sites, from Reference [1], the main advantages of the method study are the improvement in site layout and design, better utilization of manpower, equipment and materials, improvement in working approaches and safety standards, less overtime and fatigue, better workmanship, etc.

20.6 WORK MEASUREMENT

Work measurement establishes the time for a worker to perform a specific task at the set performance level. There are various applications of work measurement. The applications such as making efficiency comparison of alternative methods, balancing the tasks of team members, etc., are typical examples. The major techniques through which work measurement is accomplished are time study, work sampling, performance rating, etc. The time study technique is explored in the following sections.

20.6.1 Time Study

This approach is due to Frederick W. Taylor who formalized it in 1881. It is the commonly used technique of work measurement. The tools needed by the work study man to make sound work study analysis are a stopwatch, pencils, time study forms, an observation board, a speed indicator, a measuring tape and a calculator.

There are various reasons to select a job for time study. Some of them are as follows [1]:

(i) Worker or workers complained about the existing time standard for a specific operation.

(ii) A decision has been made to install an incentive plan; therefore standard times are required.

(iii) Compare the efficiency of two alternative approaches.

(iv) Cost of a specific job is too high.

(v) Job or operation is completely new.

(vi) Material or the working approach has been modified.

TIME STUDY APPROACH

The procedure for performing time study may vary quite signficantly from one person to another. However, the basic steps involved are shown in Figure 20.2. Steps 1, 2, 3 and 5 are self-explanatory; therefore they are not described here. The remaining steps are discussed as follows.

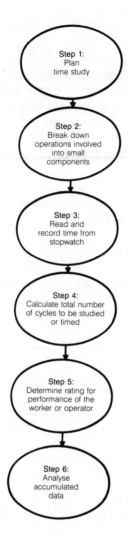

Figure 20.2.
Steps for performing time study.

STEP 4

This step is concerned with determining the number of cycles to be timed.
Thus the number of cycles, C_n, from References [4,5], is given by

$$C_n = \left\{ \frac{(k/s)\, [M\Sigma Y^2 - (\Sigma Y)^2]^{1/2}}{\Sigma Y} \right\}^2 \qquad (20.1)$$

where

k/s denotes the confidence-precision factor. [The basis for this ratio is the desired confidence level; the commonly used value is 95% (observed times within ± 2 standard errors of mean); and the precision of the element sample mean time in comparison with actual time, the usual value used, is $\pm 5\%$ precision. Thus the value of $k/s = 2/0.05 = 40$.]

M denotes the number of representative individual element times.

Y denotes the representative individual element times.

EXAMPLE 20.1

Assume that for a specific element, the values of five elemental (observation) times, in minutes, are as follows:

$$Y_1 = 10, \; Y_2 = 11, \; Y_3 = 12, \; Y_4 = 9, \; Y_5 = 13$$

If the value of the ratio $k/s = 40$, calculate the number of cycles to be timed.

Thus from the specified data we have

$$\sum_{i=1}^{5} Y_i = 55$$

and

$$\sum_{i=1}^{5} Y_i^2 = (10)^2 + (11)^2 + (12)^2 + 9^2 + (13)^2 = 615$$

Utilizing the above data in Equation (20.1) results in

$$C_n = \left\{ \frac{40[5(615) - (55)^2]^{1/2}}{55} \right\}^2$$

$$= \quad 26.45 \text{ cycles}$$

STEP 6

This step is concerned with the analysis of accumulated data. In other words, the values of select time, normal time and standard time are determined in this step.

The select time is calculated by taking the arithmetic mean of the observed times of an element in question. On the other hand, the normal time, T_n, the time an experienced worker at normal pace would take to perform an element, is given by

$$T_n = T_s \cdot F_r = T_s \cdot \text{(Performance rating in percentages)}/100 \qquad (20.2)$$

where

T_s is the select time.

F_r is the worker performance rating factor. The normal performance rating is expressed as 100%. The better-than-normal operator's pace is given by a figure above 100% and the pace slower than the normal is given by a figure less than 100%.

Lastly, the standard time is that time a work element would normally require after allowing for interruptions, for example, the allowances for unavoidable delays, fatigue, etc. Thus the standard time, T, is given by

$$T = T_n \left(1 + \frac{\text{allowance in percentages}}{100} \right) \qquad (20.3)$$

EXAMPLE 20.2

An element select time and the operator's performance rating are 0.5 minutes and 95%, respectively. Calculate the value of the normal time.
Substituting the specified data into Equation (20.2) leads to

$$T_n = 0.5 \left(\frac{95}{100} \right) = 0.475 \text{ minutes}$$

EXAMPLE 20.3

The overall operation normal time and the allowances are 0.5 minutes and 15%, respectively. Calculate the standard time.
Utilizing the given data in Equation (20.3) yields

$$T = 0.5 \left(1 + \frac{15}{100} \right) = 0.575 \text{ minutes}$$

20.7 INCENTIVE PLANS

In industry, there are several types of incentive plans used to increase workers' motivation by allowing them to earn proportionately more for their increased efforts. Thus, various types of models are used to make incentive payments. Three of them are described in the following sections [5].

20.7.1   Model I

This model is known as the straight piece-rate model. In this plan the payment is made at a constant amount per unit of output. Because of its simplicity, this method is used when making payments to harvest workers and various other types of contract workers. The hourly earnings, D, in dollars can be calculated from the following equation:

$$D = h_r \cdot T \cdot (T_a)^{-1} \tag{20.4}$$

where

T denotes the standard time to produce one piece.
T_a denotes the actual time to produce one piece.
h_r is the rate in dollars on an hourly basis ($/hour).

20.7.2   Model II

This method is known as the piece-rate with a guaranteed base wages model. In this model the earnings, D, are calculated from the following equations:

$$D = \begin{cases} h_r \text{ for } T_a \geq T \\ h_r T(T_a)^{-1} \text{ for } T_a \leq T \end{cases} \tag{20.5}$$

20.7.3   Model III

This is known as the profit-sharing incentive model. In this situation, the output above the normal (day's) output is shared with the employer. Thus the earnings are calculated by using the following equation:

$$D = h_r \left[1 + \alpha \left(\frac{T - T_a}{T_a} \right) \right] \tag{20.6}$$

where α denotes the proportion of incentive for the above normal (day's) output given to the worker.

20.7.4   Benefits and Drawbacks of Wage Incentive Plans

Like any other concept, the wage incentive plans also have their advantages and disadvantages [5]. The major advantages of the incentive plans are as follows:

(i) The method improvements are stimulated because of work studies associated with incentive plans.

(ii) The need for supervision decreases.

(iii) Labor costs per unit estimation accuracy increases.

Conversely, the main disadvantages of the incentive plans are as follows:

(i) They create more administration work.

(ii) Due to faster pace of work activities, the safety and quality may suffer.

20.8 SUMMARY

The chapter explores the various aspects of work study. The reasons for performing work study and its criticisms are briefly discussed. In addition, the general approach to performing work study, work study department, and work study engineer qualifications and qualities are described.

Method study application areas and techniques such as process charts and operation analysis are briefly discussed.

The time study component of work measurement is explored in detail. Equations to calculate normal and standard times are given.

Three different wage incentive plan model equations are given along with advantages and disadvantages of the incentive plans.

20.9 EXERCISES

1. What are the advantages of work study?
2. Discuss the similarities and the differences between method study and work measurement.
3. What are the functions of a work study engineer?
4. Describe the following terms:
   (i) Select time
   (ii) Normal time
   (iii) Standard time
   (iv) Normal performance rating
5. What are the major benefits of method study?
6. Describe the term "work sampling."
7. What are the essential tools needed by the work study man to perform his job effectively?

20.10 REFERENCES

1. Rowe, K. *Management Techniques for Civil Engineering Construction*. New York:John Wiley & Sons (1975).
2. *Introduction to Work Study*. Geneva:International Labour Office (1974).
3. Maynard, H. B., ed. *Industrial Engineering Handbook*. New York:McGraw-Hill Book Company (1971).
4. Laufer, A. C. *Operations Management*. Cincinnati:South-Western Publishing Company (1975).
5. Riggs, J. L. *Production Systems: Planning, Analysis and Control*. New York:John Wiley & Sons (1976).

Index